KU-023-734

CONTENTS

Contents

F G J Norton has published successful textbooks at both A and O level. He is the author of *Advanced Level Applied Mathematics*, and co-author with L Harwood Clarke of *Additional Applied Mathematics*, *Additional Pure Mathematics* and *Pure Mathematics at Advanced Level*. He is also the author of *Mathematics* in the Pan Study Aids GCSE series.

Mr Norton has been an examiner for the London and Cambridge boards and chief examiner in Mathematics for the Associated Examining Board. He is currently Head of Mathematics at Rugby School.

Pan Study Aids for A level include:

Advanced Biology

Advanced Chemistry

Advanced Computing Science

Advanced Economics

Advanced Mathematics

Advanced Physics

Advanced Sociology

ADVANCED MATHEMATICS

F. G. J. Norton

Tracey Hallal

Pan Books London and Sydney

First published 1982 by Pan Books Ltd
in association with Heinemann Educational Books Ltd

This new and revised edition published 1987
by Pan Books Ltd, Cavaye Place, London SW10 9PG

9 8 7 6 5 4 3 2 1

© F. G. J. Norton 1987
Text design by Peter Ward
Text illustrations by ML Design

ISBN 0 330 29429 6

Photoset by Parker Typesetting Service, Leicester
Printed in Spain by Mateu Cromo, S. A. Madrid

Contents

PREFACE TO THE
SECOND EDITION

In this edition the opportunity has been taken to add some extra questions to exercises where they have seemed needed, and in a few places to bring the notation to conform with that now agreed by the G.C.E Boards.

Rugby, 1987   F.G.J.N.

THE EXAMINATION BOARDS

The addresses given below are those from which copies of syllabuses and past examination papers may be ordered

Associated Examining Board,
Stag Hill House,
Guildford, Surrey, GU2 5XJ

University of Cambridge Local
Examinations Syndicate,
Syndicate Buildings,
1 Hills Road,
Cambridge CB1 2EU

Joint Matriculation Board,
Manchester M15 6EU

University of London School
Examinations Department
(Publication Office),
52 Gordon Square,
London WC1H 0PJ

Northern Ireland Schools
Examination Council,
Examinations Office,
Beechill House,
Beechill Road,
Belfast BT8 4RS

Oxford Delegacy of Local
Examinations,
Ewert Place,
Summertown,
Oxford OX2 7BZ

Oxford and Cambridge Schools
Examination Board,
10 Trumpington Street,
Cambridge CB2 1QB

Southern Universities Joint
Board,
Cotham Road, Bristol BS6 6DD

Welsh Joint Education
Committee,
245 Western Avenue,
Cardiff CF5 2YX

USING THIS BOOK

This book is designed to give students a thorough guide for revision before the A level examinations. It is intended to be suitable either for the student working alone or for class revision, especially where there is a wide range of ability and need in the class. Students should be able to work through on their own, or with a bare minimum of additional explanation, enabling the teacher present to give attention to individuals, while the class follows the text.

Although there is still some variation between the G.C.E Boards, they have all revised their syllabuses recently, and in many cases produced a single syllabus combining the 'traditional' and the 'modern'. In applied mathematics there is more interest in probability, applied calculus and vector methods, less in classical mechanics. This book covers all the single-subject mathematics syllabuses current at present, and most of the material in the Pure Mathematics and Further Mathematics (or 'double subject') syllabuses. It must therefore contain material that some candidates do not need to cover. All candidates should consult an up-to-date syllabus at school, college or in the local Public Library, and they are advised to cross out topics in this book that are not in the syllabus that they are following.

Booklets of past papers can be obtained from each Board (addresses are given on page 9), and should be worked through in conjunction with the exercises in this book.

Confidence in mathematics only comes from 'doing'. Each section of the book has explanatory text, worked examples, then some questions testing that particular topic. Many of the questions are short, since most G.C.E. Boards now have a paper of short questions. The longer questions are similar to those set in G.C.E papers, but, since questions have *not* been taken from past papers, candidates can attempt the past papers of their own Boards without finding that they have already answered those questions.

GENERAL EXAMINATION HINTS

1   **Read the questions carefully,** especially the numbers, signs and indices.

2   **Choose your questions carefully;** in particular, start with a question that you can do. It is most unlikely that you have to start with question 1.

3   In short-answer papers, **if you cannot see how to do a question** straight away, **leave it out for the time being,** and return later if you have time. You do not have to answer all the questions to pass the examination, or even to get quite a respectable grade!

4   Always **make a rough estimate of any calculation**. If you are using a calculator, set out each calculation clearly, so that you (and the examiner) can see what you are trying to do.

5   Always **try to keep solutions as simple as possible** – if you find a difficult method, you may make a mistake in following it. In particular, integrals are often simpler than they appear at first sight!

6   **Do not cross out an answer** because you think that it is wrong. Part of your solution may well be correct, and you will lose the marks for this if you have crossed it out.

7   If you have finished before the end of the examination, **check your work carefully**.

8   If you are 'stuck' in any one question, **check that you have used all the information given**, and see whether you can get any ideas from an earlier part of the question.

9   Where a question says, 'hence or otherwise', **it is almost invariably easier to try 'hence'**, using the hint given.

10 In examinations in which formulae sheets are provided, do make sure that you are **thoroughly familiar** with the **formulae sheets** provided.

NOTATION USED

Most of the G.C.E. Boards have now agreed on a standard notation; the following is used in this book:

$\mathbf{Z}$ — the set of integers $\{0, \pm1, \pm2, \pm3, \ldots\}$

$\mathbf{Z}^+$ — the set of positive integers $\{1, 2, 3, \ldots\}$

$\mathbf{Q}$ — the set of rationals

$\mathbf{Q}^+$ — the set of positive rationals

$\mathbf{R}$ — the set of real numbers

$\mathbf{R}^+$ — the set of positive real numbers

$\mathbf{R}_0^+$ — the set of positive real numbers and zero

$\sqrt{x}$ — the positive square root only

$|x|$ — the modulus of $x = \begin{cases} x \text{ if } x \geqslant 0, \\ -x \text{ if } x < 0 \end{cases} \quad x \, \varepsilon \, \mathbf{R}$

$f'(x), f''(x), \ldots$ — the first, second, ... derivatives of $f(x)$ with respect to x

z — a complex number $z = x + iy = r(\cos\theta + i \sin \theta)$

z^*(or $\bar{z}$) — the complex conjugate of z; $z^* = x - iy = r(\cos \theta - i \sin \theta)$

$\overrightarrow{AB}$ — the vector represented in magnitude and direction by the line segment $\overrightarrow{AB}$

$\mathbf{i, j, k}$ — unit vectors in the direction of the coordinate axes

$|\mathbf{a}|$ — the magnitude of the vector $\mathbf{a}$

$\mathbf{a.b}$ — the scalar product of vectors $\mathbf{a}$ and $\mathbf{b}$

$P(A)$ — the probability of the event A

$\binom{n}{r}$ — the binomial coefficient $\dfrac{n!}{r!(n-r)!}$ previously written ^nC_r or $_nC_r$

FUNCTIONS

CONTENTS

A **function** maps one element in a set (the domain) into **one and only one** element in another set (the range).

*The alternative notation $f°g$ is sometimes used.

The **composite** function fg* means 'first find the image under g, then the image of that under f',

e.g., if $f:x \to \sin x$ and $g:x \to 2x$,

then $fg:x \to \sin 2x$ and $gf:x \to 2 \sin x$

The **inverse of fg** is $(g^{-1})(f^{-1})$, usually written $g^{-1}f^{-1}$

An **even** function f is one such that $f(x)=f(-x)$; its graph is symmetrical about Oy. Examples of even functions are

$$f:x \to x^2, f:x \to \cos x \text{ and } f:x \to e^{|x|}$$

An **odd** function f is one such that $f(x)=-f(-x)$; its graph is symmetrical about the origin. Examples of odd functions are

$$f:x \to x, f:x \to \sin x \text{ and } f:x \to \frac{1}{x}$$

A **periodic** function f is one such that $f(x)=f(x+a)$. If a is the smallest number for which this is true, the function is said to have period a. Examples of periodic functions are

$$f:x \to \sin x, \text{ period } 2\pi, \text{ since } \sin x=\sin(x+2\pi)=\sin(x+4\pi) \ldots$$

$$f:x \to \tan x, \text{ period } \pi, \text{ since } \tan x=\tan(x+\pi)=\tan(x+2\pi) \ldots$$

FUNCTIONS

A function maps any one element x in a set (called the domain) into one and only one element (often denoted by y, the image of x) in another set (called the range). A function can map one element in the domain into one and only one element in the range (one-to-one) or can map many elements in the domain into just one element in the range (many-to-one).

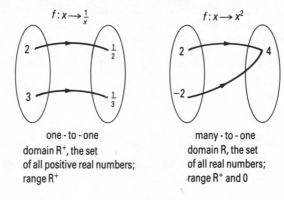

one - to - one
domain R^+, the set
of all positive real numbers;
range R^+

many - to - one
domain R, the set
of all real numbers;
range R^+ and 0

Fig 1.1

The function that maps every number into its reciprocal is written $f:x \rightarrow 1/x$ or, denoting elements in the range by y, $y=1/x$. This is a one-to-one function. By contrast, the 'square' function $f:x \rightarrow x^2$ maps every number into its square, so that, e.g. $+2$ and -2 have the same image: 4; this is an example of a many-to-one function.

The function $f:x \rightarrow \sin x$ maps $\pi/6$, $5\pi/6$, $13\pi/6$ etc. into 1/2. If the domain is the set of all real numbers then the function is many-to-one, but if the domain is restricted to $\{x:-\pi/2 \leqslant x \leqslant \pi/2\}$ then the function is one-to-one. This is not the only possible domain, for which $f:x \rightarrow \sin x$ is one-to-one. For example, the domain $\{x:0 \leqslant x \leqslant \pi/2$ or $\pi < x \leqslant 3\pi/2\}$ also makes the function one-to-one.

INVERSE FUNCTIONS

If a function f maps an element x in the domain into an element y in the range, the function f^{-1} that maps y back into x is called the inverse function. Since there must be a unique image under a function, only one-to-one functions have inverse functions. We can deduce from a graph whether an inverse function exists, for we can see whether or not one and only one value of x corresponds to any one value of y.

Example 1.1 If the function f is such that an inverse function f^{-1} exists, find a possible domain and range for each of the following:

(a) $f:x \rightarrow x^2$ (b) $f:x \rightarrow \sin x$ (c) $f:x \rightarrow \cos x$

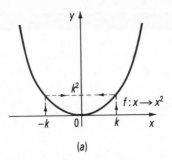
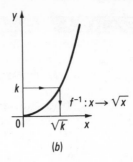

Fig 1.2

(a)   (b)

Figure 1.2(a) shows the graph of $y=x^2$. For every real value of x, there is an image y, but for every positive value of y there are two possible images, one positive and one negative (the non-positive number 0 has only one image). If, however, we define the domain as the set of all non-negative numbers then the range is the set of all non-negative numbers (Fig 1.2(b)) and the function is one-to-one, and has an inverse.

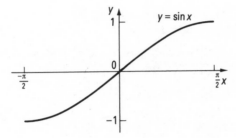

Fig 1.3

We can see from Fig 1.3 that if the domain is $\{x: -\pi/2 \leqslant x \leqslant \pi/2\}$ and the range the set $\{y: -1 \leqslant y \leqslant 1\}$, then the function $f:x \to \sin x$ has an inverse. The values for x in this case are called the **principal values**. One can see the restrictions on the domain that are built into any calculator; use of, for example, the inv sin button produces one and only one image because the domain is restricted by the calculator.

Considering the function $f:x \to \cos x$, Fig 1.4 shows that if the domain is $0 \leqslant x \leqslant \pi$ and the range $-1 \leqslant y \leqslant 1$, then the function is one-to-one and an inverse function exists.

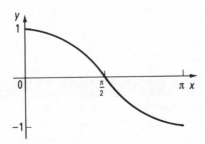

Fig 1.4

COMPOSITE FUNCTIONS

If $f:x \rightarrow 2x$ and $g:x \rightarrow \sin x$, $gf(\pi/6)$ means that we first find the image of $\pi/6$ under f, i.e. $f(\pi/6)=\pi/3$, then the image of $\pi/3$ under g, i.e. $g(\pi/3)=\sin(\pi/3)=\frac{1}{2}\sqrt{3}$. This is not the same as $fg(\pi/6)$, for $fg(\pi/6)=f(1/2)=1$, since $g(\pi/6)=\sin \pi/6=1/2$ and $f(1/2)=2\times 1/2=1$. To find the inverse of this composite function we must find first the image under g^{-1}, then under f^{-1}, e.g. $f^{-1}g^{-1}(1/2)=f^{-1}(\pi/6)=\pi/12$. Thus $(gf)^{-1}=f^{-1}g^{-1}$.

N.B. **$gf(x)$ means 'first find the image of x under f, then the image of $f(x)$ under g'.**

Example 1.2 If $f:x \rightarrow 1+x$ and $g:x \rightarrow 1/x$, find

(a) $gf(1/2)$ (b) $fg(1/2)$

(c) $f^{-1}g^{-1}(2/3)$ (d) $g^{-1}f^{-1}(2/3)$

(a) $f(1/2)=1+1/2=3/2$ and $g(3/2)=1/(3/2)=2/3$

(b) $g(1/2)=1/(1/2)=2$ and $f(2)=3$
       $\therefore fg(1/2)=3$

(c) $g^{-1}(2/3)=1/(2/3)=3/2$ and $f^{-1}(3/2)=3/2-1=1/2$
       $\therefore f^{-1}g^{-1}(2/3)=1/2$

(d) $g^{-1}f^{-1}(2/3)=g^{-1}(-1/3)=-3$

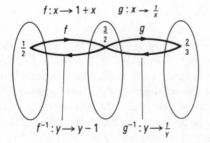

Fig 1.5

The composite function gf can be written as a single function h where $h:x \rightarrow 1/(1+x)$, whereas fg is the single function H where

$$H:x \rightarrow \frac{1}{x}+1$$

EVEN AND ODD FUNCTIONS

If $f(x)$ denotes the image under f of x, a function f such that $f(-x)=f(x)$ for all x is called an **even** function. The graph of an even function is symmetrical about the line $x=0$. Examples of even functions are $f:x \rightarrow 1-x^2$; $f:x \rightarrow \cos x$; $f:x \rightarrow e^{-x^2}$ (Fig 1.6).

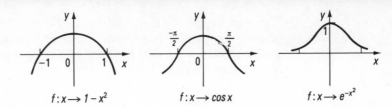

Fig 1.6

A function f such that $f(-x)=-f(x)$ for all x is called an **odd** function. The graph of an odd function is symmetrical about the origin. Examples of odd functions are $f:x \rightarrow x$; $f:x \rightarrow x^3$ and $f:x \rightarrow \sin x$ (Fig 1.7).

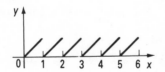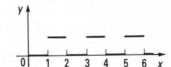

Fig 1.7

PERIODIC FUNCTIONS

A function f such that $f(x+a)=f(x)$ for **all** x is said to be periodic, period a. The most important periodic functions are $f:x \rightarrow \sin x$ and $f:x \rightarrow \cos x$ (period 2π), $f:x \rightarrow \tan x$ (period π). Figure 1.8 shows two other periodic functions.

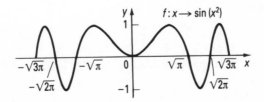

Fig 1.8

Example 1.3 Find whether the function $f:x \rightarrow \sin(x^2)$ is even or odd, and whether or not it is periodic.

Since $f(-x)=\sin(-x)^2=\sin(x^2)=f(x)$, the function f is even, although the sin function is odd.

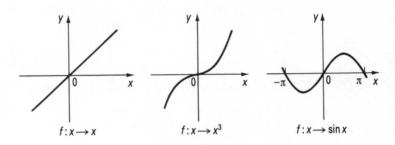

Fig 1.9

To see whether the function is periodic, we know the sin function is periodic, period 2π, but unequal intervals of x are required to give equal intervals of x^2, so this function is not periodic. These properties are illustrated in Fig 1.9

QUESTIONS

1   If the domain and range are both **R**, the set of all real numbers, which of the following functions are one-one and which many-one?

   (a) $f:x \to x^3$ (b) $f:x \to x^4$
   (c) $f:x \to (x-1)^2$ (d) $f:x \to \sin^2 x$

2   Give a possible domain and range for each of the following functions if it is to have an inverse function:

   (a) $f:x \to (x-1)^2$ (b) $f:x \to (x+2)^4$
   (c) $f:x \to \sin^2 x$ (d) $f:x \to 1/x^2$

3   If $h=fg$ and $H=gf$, find h^{-1} and H^{-1} when

   (a) $f:x \to 3x$ and $g:x \to x+2$
   (b) $f:x \to x^2$ and $g:x \to x+1$
   (c) $f:x \to \sin x$ and $g:x \to x^2$

4   Sketch the function f defined by

$$f:x \begin{cases} -1 \text{ if } x \leqslant -1 \\ x \text{ if } -1 < x \leqslant 1 \\ 1 \text{ if } 1 < x \end{cases}$$

   (a)   Is there an inverse function?
   (b)   Is f even or odd or neither?

5   If the domain is $\mathbf{R}_0^+$ the set of all non-negative real numbers, and $f:x \to e^x$, find the range and the inverse function f^{-1} if it exists.

6   If the domain is $\{x:-1 \leqslant x \leqslant 1\}$ and $f:x \to x\sqrt{(1-x^2)}$, sketch f to show that there is no inverse function.

7   If $f:x \to \sin x$ and $g:x \to 2x$ find single functions h and H such that $h=gf$ and $H=fg$. Find also h^{-1} and H^{-1}.

8   A function f is periodic, period 2, and is defined by

$$f:x \to \begin{cases} x-2k & \text{if } 2k \leqslant x < 2k+1 \\ 2k+2-x & \text{if } 2k+1 \leqslant x < 2k+2 \end{cases} \text{ where } k \text{ is any integer}$$

   Sketch f if the domain is the set **R** of all real numbers. Is f even or odd?

9   Are the following functions even or odd? Find their period if they are periodic.

   (a) $f:x \to e^x$ (b) $f:x \to e^{x^2}$
   (c) $f:x \to 1/x$ (d) $f:x \to \sin 2x$
   (e) $f:x \to \sin(x-\pi/2)$ (f) $f:x \to \cos(\pi-x)$

THE QUADRATIC FUNCTION

CONTENTS

NOTES

The quadratic equation $ax^2 + bx + c = 0$ has

and

two real distinct roots if $b^2 > 4ac$
equal roots if $b^2 = 4ac$
no real roots if $b^2 < 4ac$

The sum of the roots is $-b/a$: the product of the roots is c/a.
The graph is concave upwards if a is positive,
downwards if a is a negative.

a positive   a negative

Fig 2.1

THE QUADRATIC FUNCTION

Since

$$ax^2 + bx + c = a\left(x^2 + \frac{bx}{a}\right) + c$$

$$= a\left(x + \frac{b}{2a}\right)^2 + c - \frac{b^2}{4a},$$

$$ax^2 + bx + c = 0 \text{ only if } \frac{b^2}{4a} \geqslant c; \text{ i.e. } b^2 - 4ac \geqslant 0.$$

The expression $b^2 - 4ac$ is called the discriminant of the quadratic. If $b^2 - 4ac > 0$, the quadratic equation has two real distinct roots; if $b^2 - 4ac = 0$ the quadratic equation has two equal roots; and if $b^2 - 4ac < 0$ the quadratic equation has no real roots.

If

$$ax^2 + bx + c = 0$$

$$\left(x + \frac{b}{2a}\right)^2 = \frac{b^2 - 4ac}{4a^2}$$

i.e.

$$x = \frac{-b \pm \sqrt{b^2 - 4ac}}{2a}$$

a formula which enables us to solve any quadratic equation.

Example 2.1 Find the range of values of p if $px^2+x+p=0$ has real roots.

Comparing $px^2+x+p=0$

with $ax^2+bx+c=0$

$a=p$, $b=1$ and $c=p$

$\therefore b^2-4ac\geqslant 0 \Rightarrow 1-4p^2\geqslant 0,$

i.e. $p^2\leqslant \dfrac{1}{4}$

$-\dfrac{1}{2}\leqslant p\leqslant \dfrac{1}{2}$

MAXIMUM AND MINIMUM VALUES OF THE QUADRATIC FUNCTION

Writing ax^2+bx+c as $a(x+b/2a)^2+c-b^2/4a$, we see that if a is positive, $ax^2+bx+c\geqslant c-b^2/4a$

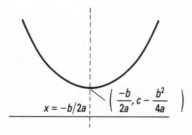

Fig 2.2

equality occurring only when $x=-b/2a$; if a is negative, then $ax^2+bx+c\leqslant c-b^2/4a$, equality again occurring only when $x=-b/2a$.

Example 2.2 Find the least value of $2x^2+12x+7$.

$$2x^2+12x+7= \quad 2(x^2+6x)+7$$
$$= \quad 2(x+3)^2-18+7$$
$$= \quad 2(x+3)^2-11$$

Since $(x+3)^2$ is a perfect square, it is never negative. Therefore, $2x^2+12x+7\geqslant -11$, the least value, -11, occurring only when $x=-3$.

SOLUTION OF TWO SIMULTANEOUS EQUATIONS ONE LINEAR AND ONE QUADRATIC

To solve a pair of simultaneous equations, one of which is linear and the other quadratic, find one variable in terms of the other from the linear equation and substitute in the quadratic.

Example 2.3 Solve the simultaneous equations

$$x+2y=3 \tag{1}$$
$$x^2+3y^2=13 \tag{2}$$

From equation (1), $\qquad x = 3 - 2y \qquad$ (3)

Substituting in (2), $\qquad (3-2y)^2 + 3y^2 = 13$

$\therefore \qquad\qquad 9 - 12y + 4y^2 + 3y^2 = 13$

i.e. $\qquad\qquad 7y^2 - 12y - 4 = 0$

i.e. $\qquad\qquad (y-2)(7y+2) = 0$

$\qquad\qquad\qquad y = 2 \text{ or } -2/7.$

Substituting $y = 2$ in (3), $\qquad\qquad x = -1$

Substituting $y = -2/7$ in (3), $\qquad\qquad x = 3\frac{4}{7}$

The solutions are

$\qquad x = -1 \text{ and } y = 2$

or $\qquad x = 3\frac{4}{7} \text{ and } y = -\frac{2}{7}.$

N.B. Take care to pair the solutions correctly; it is not sufficient to say just

$$x = -1 \text{ or } 3\tfrac{4}{7}, \quad y = 2 \text{ or } -\tfrac{2}{7}$$

SYMMETRIC FUNCTIONS OF THE ROOTS OF A QUADRATIC EQUATION

If the roots of $ax^2 + bx + c = 0$ are $x = \alpha$ and $x = \beta$,

$$ax^2 + bx + c \equiv a(x-\alpha)(x-\beta)$$
$$\equiv a(x^2 - (\alpha+\beta)x + \alpha\beta)$$

Equating the coefficients of x, $b = -a(\alpha+\beta)$, i.e. $(\alpha+\beta) = -b/a$, equating constants, $c = a\alpha\beta$, i.e. $\alpha\beta = c/a$, thus

the sum of the roots of a quadratic is $-b/a$

the product of the roots is c/a.

Example 2.4 If the roots of $2x^2 - 5x + 1 = 0$ are α and β, find the equation whose roots are $\alpha^2 + 1$, $\beta^2 + 1$.

Comparing $\qquad 2x^2 - 5x + 1 = 0$

with $\qquad\qquad ax^2 + bx + c = 0,$

we see $\qquad\qquad\qquad \alpha + \beta = \dfrac{5}{2}$

$$\alpha\beta = \dfrac{1}{2}$$

Now $\qquad\qquad\qquad \alpha^2 + \beta^2 = (\alpha+\beta)^2 - 2\alpha\beta$

$$= \left(\frac{5}{2}\right)^2 - 2\left(\frac{1}{2}\right)$$

$$= \frac{21}{4}$$

$\therefore \qquad\qquad (\alpha^2+1) + (\beta^2+1) = \dfrac{21}{4} + 2 = \dfrac{29}{4},$

and $(\alpha^2+1)(\beta^2+1)$ $\qquad = \alpha^2\beta^2+\alpha^2+\beta^2+1$

$$= \left(\frac{1}{2}\right)^2 + \frac{21}{4} + 1 = \frac{26}{4}$$

The required equation is

$$x^2 - \frac{29}{4}x + \frac{26}{4} = 0$$

i.e. $4x^2 - 29x + 26 = 0$

---

1   Find the range of values of k if each of the following equations has two real roots.

   (a) $x^2+2x+k=0$ (b) $3x^2-2x+k=0$
   (c) $x^2-kx+1=0$ (d) $x^2+3x+1=k$

2   If $x \in \mathbf{R}$, find the greatest value of
   (a) $4-x^2$ (b) $4-(x-1)^2$
   (c) $4+2x-x^2$ (d) $3+4x-x^2$

Write down in each case the value of x which gives the greatest value for the expression.

3   If $x \in \mathbf{R}$, find the least value of each of the following expressions, and the value of x which gives that least value.

   (a) x^2+4 (b) $(x+1)^2+3$
   (c) x^2+2x+4 (d) x^2+4x+3

4   Solve for x and y

   (a) $3x+y=5,\ x^2+y^2=5$
   (b) $3x+y=5,\ x^2+2xy+y^2=9$
   (c) $3x+y=5,\ xy=2$

5   Show that there are no real solutions to the equations
$$2x+y=4,\ 4x^2+y^2=4$$

6   Show that there are no real solutions to the equations
$$x+y=4,\ x^2+y^2=1$$
and illustrate this statement by a geometrical sketch.

7   Find the range of value of k for which there are solutions of
$$x+y=k,\ x^2+y^2=4$$
and illustrate your result geometrically.

8   If α and β are the roots of $x^2+4x+1=0$, form the equation whose roots are $\alpha+1$, $\beta+1$.

9   Form the equation whose roots are each one less than a root of the equation $x^2-5x-2=0$.

10   If the roots of $3x^2+4x-2=0$ are α and β, form the equation whose roots are $1/\alpha$ and $1/\beta$.

11 Form the equation whose roots are each the reciprocal of a root of $5x^2-4x-2=0$.

12 If α and β are the roots of $x^2-5x-1=0$, form the equation whose roots are α^2+1 and β^2+1.

13 Sketch the graph of $y=x^2-2x+2$ to show the least value of y if $x\epsilon\mathbf{R}$. If x is restricted so that $x>2$, what is now the least value of y?

14 Find the range of values of k if the equation $x^2-2kx+k=0$ has real roots.

15 Find the values of k if the equation $x^2-2kx+k^2=1$ has real roots.

16 Find the values of k for which the equation $kx^2+2x+3=0$ has roots differing by 4.

17 Find the range of values of x for which $x+2y=1$ and $x^2-4y^2\leqslant 5$.

18 Solve simultaneously $2x+y=4$ and $2x^2=xy-1$.

19 The roots of the quadratic equation $x^2+px+q=0$ are α and β. If $\alpha-\beta=1$ and $\alpha^2+\beta^2=41$, find p and q.

20 The roots of the quadratic equation $x^2-3x+1=0$ are α and β. Find the quadratic equation whose roots are α^3/β and β^3/α.

21 If the roots of the equation $x^2+ax+b=0$ are α and $k\alpha$, show that $(k+1)^2b=ka^2$.

22 Given that the equations $x^2+ax+b=0$ and $x^2+3ax+4b=0$ have a common root, show that $9b=2a^2$.

THE REMAINDER THEOREM, INDICES AND LOGARITHMS

CONTENTS

NOTES

REMAINDER THEOREM

When a polynomial $f(x)$ is divided by $(x-a)$, the **remainder** is $f(a)$; if $f(a)=0$, $(x-a)$ is a **factor** of $f(x)$.

If $(x-a)^2$ is a factor of $f(x)$, so that $(x-a)$ is a **repeated factor**, $f(a)=0$ and $f'(a)=0$.

INDICES

The laws of **indices** are

$$a^m \times a^n = a^{m+n}, \ a^m \div a^n = a^{m-n} \text{ and } (a^m)^n = a^{mn}$$

From these we deduce

$$a^0 = 1, \ a^{-m} = 1/a^m \text{ and } a^{1/m} = \sqrt[m]{a}, \text{ especially } a^{1/2} = \sqrt{a}$$

LOGARITHMS

By definition of the log function, $\boldsymbol{y = a^x \Longleftrightarrow x = \log_a y}$

e.g.   since $100 = 10^2$, $\log_{10} 100 = 2$

From the laws of indices, we deduce

$\log x + \log y = \log(xy)$; $\log x - \log y = \log(x/y)$; $\log(x^n) = n\log x$

LINES OF BEST FIT

If there is a relation of the form $y = ax^2$, plot y against x^2.
If there is a relation of the form $y = a/x$, plot y against $1/x$.

If there is a relation of the form $y = ax^n$, where n is unknown, write this as $\log y = \log a + n\log x$, and plot $\log y$ against $\log x$.

THE REMAINDER THEOREM

When a polynomial $f(x)$ is divided by a linear expression, say $x-a$, the remainder R will be a constant, for if there are any terms in x in the remainder we still can divide by $(x-a)$. Thus

$$f(x) \equiv (x-a)\,Q(x) + R$$

where $Q(x)$ is the quotient. Since this is an identity it is true for all values of x, so substitute $x=a$,

i.e. $\qquad f(a)=0\times Q(a)+R$

$\therefore \qquad\qquad R=f(a)$

This result is most useful for finding factors of a polynomial, for if $(x-a)$ is a factor the remainder is zero, i.e. $f(a)=0$.

Example 3.1 Find the linear factors of $x^3-6x^2+3x+10$

Try the factors of the constant term, $\pm 1, \pm 2, \pm 5$.
Denoting $x^3-6x^2+3x+10$ by $f(x)$,

$$f(1)=(1)^3-6(1)^2+3(1)+10\neq 0$$
$\therefore \qquad$ $(x-1)$ is not a factor
$$f(-1)=(-1)^3-6(-1)^2+3(-1)+10=0$$
$\therefore \qquad$ $(x+1)$ is a factor
Similarly $\qquad f(2)=0$, so $(x-2)$ is a factor
and $\qquad\qquad f(5)=0$, so $(x-5)$ is a factor

Since $f(x)$ is a cubic there are only three linear factors, and since the coefficient of x^3 is 1,

$$x^3-6x^2+3x+10\equiv(x+1)(x-2)(x-5)$$

REPEATED FACTORS

If $(x-a)^2$ is a factor of $f(x)$,

$$f(x)\equiv(x-a)^2\,Q(x)$$

Let $f'(x)$ denote the derived function, i.e. $f'(x)\equiv df/dx$,
$\therefore \qquad f'(x)\equiv(x-a)^2Q'(x)+2(x-a)Q(x)$
$$\equiv(x-a)[(x-a)Q'(x)+2Q(x)]$$
$\therefore \qquad$ $(x-a)$ is a factor of $f'(x)$
thus not only is $f(a)=0$, but also $f'(a)=0$.

Example 3.2 Find the values of constants A and B if $(x+1)^2$ is a factor of $2x^4+7x^3+6x^2+Ax+B$

If $\qquad f(x)=2x^4+7x^3+6x^2+Ax+B$
$\qquad\quad f'(x)=8x^3+21x^2+12x+A$

Since $(x+1)$ is a factor of $f(x)$, $f(-1)=0$

i.e. $\qquad 2(-1)^4+7(-1)^3+6(-1)^2+A(-1)+B=0$
$\qquad\qquad A-B=1 \qquad\qquad\qquad\qquad\qquad\qquad (1)$

Since $(x+1)$ is also a factor of $f'(x)$, $f'(-1)=0$

i.e. $\qquad 8(-1)^3+21(-1)^2+12(-1)+A=0$
$\qquad\qquad A=-1$

Substituting $A=-1$ in (1), we have $B=-2$.

If required, we can use the factor theorem to show that $(x+2)$ is a factor of $f(x)$ and also that $(x-\frac{1}{2})$ is a factor of $f(x)$. Since the coefficient of x^4 in $f(x)$ is 2,

$$2x^4+7x^3+6x^2-x-2\equiv2(x+1)^2(x+2)(x-\tfrac{1}{2})$$
$$\equiv(x+1)^2(x+2)(2x-1)$$

INDICES

The laws of indices are

$$a^m\times a^n=a^{m+n},\ a^m\div a^n=a^{m-n}\text{ and }(a^m)^n=a^{mn}$$

From these it can be deduced that

$$a^0=1,\ a^{-m}=\frac{1}{a^m}\text{ and }a^{1/m}=\sqrt[m]{a}$$

Negative and fractional indices are particularly important in calculus.

Example 3.3
(a) $x^{1/2}=\sqrt{x}$

(b) $x^{-1/2}=\dfrac{1}{\sqrt{x}}$

(c) $x^{-2}=\dfrac{1}{x^2}$ and

(d) $(x+1)^{-1}=\dfrac{1}{x+1}$

Example 3.4
(a) $\displaystyle\int 2\sqrt{x}\,dx=\int 2x^{1/2}dx=\frac{4}{3}x^{3/2}+C$

(b) $\displaystyle\int\frac{2}{x^2}dx=\int 2x^{-2}dx=-2x^{-1}+C=\frac{-2}{x}+C$

(c) $\displaystyle\int\frac{2}{x^{1/2}}dx=\int 2x^{-1/2}dx=4x^{1/2}+C$

Example 3.5
(a) $\displaystyle\int_1^4 x^{1/2}dx=\left[\frac{2}{3}x^{3/2}\right]_1^4$

$$=\frac{2}{3}(4)^{3/2}-\frac{2}{3}(1)^{3/2}$$

$$=\frac{2}{3}\Big[8-1\Big]$$

$$=\frac{14}{3}$$

(b) $\int_{2}^{3} x^{-2}dx = \left[-x^{-1} \right]_{2}^{3} = -\frac{1}{3} + \frac{1}{2} = \frac{1}{6}$

RATIONALIZING THE DENOMINATOR

Since $(x-y)(x+y) = x^2 - y^2$,

$$\frac{1}{\sqrt{2}+1} = \frac{(\sqrt{2}-1)}{(\sqrt{2}+1)(\sqrt{2}-1)}$$

$$= \frac{\sqrt{2}-1}{2-1}$$

$$= \sqrt{2}-1$$

This sometimes helps in evaluating expressions containing surds, as below, but more important, helps when manipulating complex numbers.

Example 3.6 Find the sum of

$$\frac{1}{\sqrt{2}+1} + \frac{1}{\sqrt{3}+\sqrt{2}} + \frac{1}{\sqrt{4}+\sqrt{3}} \cdots \frac{1}{\sqrt{100}+\sqrt{99}}$$

Using the result above, we can see that

$$\frac{1}{\sqrt{3}+\sqrt{2}} = \frac{(\sqrt{3}-\sqrt{2})}{(\sqrt{3}+\sqrt{2})(\sqrt{3}-\sqrt{2})} = \sqrt{3}-\sqrt{2}$$

$\therefore$ The sum of the series is

$$(\sqrt{2}-1) + (\sqrt{3}-\sqrt{2}) + (\sqrt{4}-\sqrt{3}) \ldots (\sqrt{100}-\sqrt{99})$$
$$= \sqrt{100} - 1 = 9$$

LOGARITHMS

The logarithmic function can be defined by saying if $y = a^x$, then $x = \log_a y$

For example,   since $1000 = 10^3$, $\log_{10} 1000 = 3$; since $0.1 = 10^{-1}$, $\log_{10} 0.1 = -1$

From the laws of indices we can deduce

$$\log_a x + \log_a y = \log_a (xy); \log_a x - \log_a y = \log_a \left(\frac{x}{y} \right)$$

and $\log_a (x^n) = n \log_a x$

The Napierian or natural logarithm $\log_e x$ is usually written $\ln x$; $\log_{10} x$ is usually written $\lg x$.

Example 3.7 Find the value of (a) $\log_2 \sqrt[3]{2}$, (b) $\log_4 \sqrt{2}$

(a)   Express $\sqrt[3]{2}$ as a power of 2.

Since $\sqrt[3]{2}=2^{1/3}$ and $\log_a(a^n)=n$

$$\log_2(2^{1/3})=\frac{1}{3}$$

(b)   Express $\sqrt{2}$ as a power of 4.

Since $\sqrt{2}=2^{1/2}=4^{1/4}$

$$\log_4\sqrt{2}=\log_4(4)^{1/4}=\tfrac{1}{4}$$

Example 3.8 Express $\log_{10}x$ as a multiple of $\log_e x$

If $\log_{10}x=N$, $10^N=x$,

i.e. $\log_e(10^N)=\log_e x$

i.e. $N\log_e10=\log_e x$

$$\log_{10}x=\frac{1}{\log_e 10}\log_e x$$

This can of course be written

$$\lg x=\frac{\ln x}{\ln 10}$$

USE OF LOGARITHMS TO DETERMINE THE RELATION BETWEEN TWO VARIABLES

If two variables x and y are thought to be related by a law of the form $y=mx+c$, then points corresponding to pairs of values of x and y should lie on a straight line. If the relation between the variables is of the form $y=ax^n$, we can deduce that $\log y=\log a+n\log x$, so if $\log y$ is plotted against $\log x$, we should have a straight line graph whose intercept is $\log a$ and whose gradient is n.

Example 3.9 A certain company was started in 1950, and its annual profits at five-yearly intervals are given below.

1950	£250
1955	£1200
1960	£7000
1965	£40 000
1970	£105 000
1975	£1 000 000

If the annual profits are £y and the number of years after 1950 is n, is there a relation of the form $y=ax^n$?

If $y=ax^n$, $\log y=\log a+n\log x$. Here we know y and n, so plot $\log y$ against n. Figure 3.1 shows that all the points except one lie close to a straight line, so that apart from 1970, there does seem to be a relation of the form $y=ax^n$ between the annual profits and the number of years since the start of the company.

Since n is the number of years after 1950, the table of values is:

n	0	5	10	15	20	25
y	250	1200	7000	40 000	105 000	10^6
log y	2.4	3.1	3.8	4.6	5.0	6

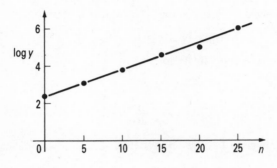

Fig 3.1

EQUATIONS CONTAINING SURDS

An equation that contains surds can often be expressed as a polynomial by squaring both sides of the equation, though we have to remember that the square of $+\sqrt{(x+a)}$ is also the square of $-\sqrt{(x+a)}$, so that the roots we obtain may not be the roots of the original equation. *Check that the roots obtained do satisfy the equation BEFORE it was squared.*

Example 3.10 Solve $x+\sqrt{x+1}=5$.

We must have the term containing the square root alone on one side of the equation, so rewrite it as

$$\sqrt{x+1}=5-x$$

Squaring both sides,
$$x+1=(5-x)^2$$
$$=25-10x+x^2$$

i.e.
$$x^2-11x+24=0$$
$$(x-3)(x-8)=0$$
$$x=3 \text{ or } 8$$

Checking, when $x=3$, $x+\sqrt{x+1}=3+\sqrt{4}=5$, so that $x=3$ is a a root of the original equation, but when $x=8$, $x+\sqrt{x+1}=8+\sqrt{9}\neq 5$, so that $x=8$ is NOT a root of the original equation. It can easily be seen that $x=8$ a root of $x-\sqrt{x+1}=5$. Thus the only root of $x+\sqrt{x+1}=5$ is $x=3$.

QUESTIONS

1. Find the linear factors of $x^3-10x^2+23x-14=0$.
2. If $f(x)=x^3-13x^2+40x-36$, find $f'(x)$. Obtain the linear factors of $f'(x)$, and verify that one of them is a repeated factor of $f(x)$.
3. Find the value of k if $(x-2)$ is a factor of $x^3-3x^2+kx-10$.
4. Find the values of A and B if $(x-2)^2$ is a factor of
   $$x^4-11x^2+Ax+B$$
5. Show that $(x+2)$ is a factor of $6x^4+11x^3-13x^2-16x+12$, and hence solve the equation
   $$6x^4+11x^3-13x^2-16x+12=0$$
6. When $x=64$, evaluate
   - (a) $x^{1/2}$ (b) $x^{1/3}$ (c) x^{-2}
   - (d) $(x^{2/3})^{-1}$ (e) $(x^{-1})^{1/6}$
7. Express in index form
   - (a) $\sqrt[4]{x}$ (b) $\dfrac{1}{x^3}$ (c) $\sqrt[3]{x^2}$
   - (d) $\sqrt[2]{x^3}$ (e) $\sqrt[3]{x^6}$
8. Integrate with respect to x
   - (a) $\sqrt[4]{x}$ (b) $\sqrt[3]{x}$ (c) $\dfrac{1}{\sqrt{x}}$
   - (d) $\dfrac{1}{\sqrt[3]{x}}$ (e) $\sqrt[3]{x^2}$
9. Find
   $$\int \frac{1}{\sqrt{(x+1)}-\sqrt{x}}\,dx$$
10. Find the value of
    - (a) $\log_2 8$ (b) $\log_8 2$
11. Find the value of a and b if
    - (a) $\log_a 3=27$ (b) $\log_b 2=8$.
12. Express $\log_5 x$ as a multiple of $\log_{10} x$. Hence find $\log_5 2$.
13. Find
    - (a) $\log_3 9$ (b) $\log_3 8$
14. Find the value of the constants A and B if $x-1$ and $x-3$ are factors of $2x^4-11x^3+Ax^2+Bx-6$.
15. Find the values of the constants A and B if x^2-5x+6 is a factor of $x^4-6x^3+10x^2+Ax+B$.
16. Find the value of the constants A and B if $(x-2)^2$ is a factor of $2x^4+Ax^3+11x^2+Bx-4$
17. If $x=16$ and $y=\frac{1}{25}$, find the value of
    - (a) $x^{1/2}$ (b) $x^{-1/2}$ (c) $xy^{1/2}$
    - (d) y^{-2} (e) $x^{1/2}(x+1)$ (f) $x^{-2}(x+9)^{1/2}$
18. Find
    - (a) $\displaystyle\int x^{1/2}(x+1)\,dx$ (b) $\displaystyle\int \frac{x+1}{x^{1/2}}\,dx$
    - (c) $\displaystyle\int \frac{x+1}{x^2}\,dx$ (d) $\displaystyle\int \frac{x+1}{x}\,dx$

19 Find (*a*) $\log_2\sqrt{8}$ (*b*) $\log_8\sqrt{2}$
        (*c*) $\log_4(2\sqrt{2})$ (*d*) $\log_2\sqrt{2}(4)$

20 Which of the following statements are true for all values of *a*, *x* and *y*?
        (*a*) $y=a^x\Longleftrightarrow\ln y=x\ln a$ (*b*) $y=a^x\Longleftrightarrow\log_a y=x$
        (*c*) $y=(a^x)^2\Longleftrightarrow\ln y=2x\ln a$

21 If $y=10^x$, which of the following would give a straight line graph?
        (*a*)   Plotting *y* against $\log x$.
        (*b*)   Plotting $\log y$ against $\log x$.
        (*c*)   Plotting $\log y$ against *x*.

22 Solve the equation
        $$2x-\sqrt{(3x+1)}=6$$

23 Solve the equation
        $$2\sqrt{2x}+\sqrt{5x-1}=7$$

INDUCTION

CONTENTS

NOTES

To prove a result for **all** values of n, first suppose that it is true for **some one** specific value of n, say $n=k$,

e.g. to prove $\qquad \sum_{r=1}^{n} r = \frac{1}{2}n(n+1)$

if this is true for some **one** value of n, say $n=k$

$$\sum_{r=1}^{k} r \equiv 1+2+3\ldots+k = \frac{1}{2}k(k+1)$$

Then the sum of the first $(k+1)$ integers, i.e. $\sum_{1}^{k+1} r$, is found by adding the $(k+1)$th integer on to what we think is the sum of the first k,

i.e. $\qquad \sum_{1}^{k+1} r = \frac{1}{2}k(k+1)+(k+1)$

$$= \frac{1}{2}(k+1)(k+2)$$

which we expect.

But the sum of the first one integer(s) is 1, which equals $\frac{1}{2} \times 1 \times 1(1+1)$, so it is true when $n=1$.

Since it is true when $n=1$, it is true when $n=2$.
Since it is true when $n=2$...
So it is true for **all positive integral** values of n

INDUCTION

$$\begin{aligned}
1 \quad &= 1 = 1^2 \\
1+3 \quad &= 4 = 2^2 \\
1+3+5 \quad &= 9 = 3^2 \\
1+3+5+7 \quad &= 16 = 4^2
\end{aligned}$$

It looks possible that the sum of the first n odd integers is always equal to n^2.

The method of induction first requires that we have an idea of the result we wish to prove, and this will be given in an examination, of course. In this example, we guessed a possible conclusion from a few special cases.

The first n odd integers are 1, 3, 5,... $(2n-1)$, and the next odd integer, the $(n+1)$th, is $(2n+1)$. If our guess is correct, the sum of the first 10 odd integers is 100; the 11th odd integer is 21, so that the sum of the first 11 odd integers is 100+21, i.e. 121, which is 11^2, so that if our guess was correct for the first ten odd integers, it is also correct for the first eleven odd integers.

Suppose that the guess is correct for the first k odd integers, i.e. $1+3+5...(2k-1)=k^2$. Then the sum of the first $(k+1)$ odd integers is

$$1+3+5+7...+(2k-1)+(2k+1)$$
$$=k^2 \qquad\quad + 2k+1$$
$$=(k+1)^2$$

so that if the guess was correct for the first k odd integers, it is also correct for the first $(k+1)$ odd integers. But the 'sum' of the first one odd integer(s) is 1, i.e. 1^2, so the guess is true for the first one odd integer; therefore it is true for the first two odd integers. Since it is true for the first two odd integers, it is true for the first three odd integers; and so on for all positive integral values of n.

There are two essential stages of the proof by induction.

1 Suppose that the result is true for some one specific value of n, and carry out the operation (often adding the next term) to see whether it is true for the next value of n.

2 Find a starting value, e.g. the sum of the first two terms, the expansion of $(x+a)^2$, the angle-sum of a three-sided polygon (a triangle).

Example 4.1 Prove by induction that

$$\sum_{r=1}^{n} r^3 \equiv 1^3+2^3+3^3...+n^3=[\tfrac{1}{2}n(n+1)]^2$$

Suppose this is true for some one specific value of n, and call this value k. Then the sum of the first k terms is

$$1^3+2^3+3^3...+k^3=[\tfrac{1}{2}k(k+1)]^2$$

In this case, the sum of the first $(k+1)$ terms will be

$$1^3+2^3+3^3...+k^3+(k+1)^3$$
$$=[\tfrac{1}{2}k(k+1)]^2+(k+1)^3$$
$$=\tfrac{1}{4}(k+1)^2[k^2+4(k+1)]$$
$$=\tfrac{1}{4}(k+1)^2(k^2+4k+4)$$
$$=\tfrac{1}{4}(k+1)^2(k+2)^2$$

which we expected, since this is the result obtained by writing $(k+1)$ in place of n in the formula. Thus if the formula is true for some one value of n, it will be true for the 'next' value of n. But $1^3=1=\tfrac{1}{4}(1)^2(2)^2$ so the formula is true when $n=1$, therefore it is true when $n=2$. Since it is true when $n=2$, it is true when $n=3$; since it is true when $n=3$, it is true when $n=4$, and so on for all positive integral values of n.

Example 4.2 Prove by induction that all numbers of the form $n(n+1)(n+2)$ are multiples of 6.

It often helps if we familiarize ourselves with the problem by trying simple numerical values of n first; here, when $n=1$, the number is $1\times2\times3=6$; when $n=2$, the number is $2\times3\times4=24$, a multiple of 6.

Suppose it is true for some one value of n, say $n=k$, so that $k(k+1)(k+2)$ is a multiple of 6, i.e. $k(k+1)(k+2)=6m$ where m is an integer. Then we want to show that $(k+1)(k+2)(k+3)$ is also a multiple of 6. Now

$$(k+1)(k+2)(k+3)=k(k+1)(k+2)+3(k+1)(k+2)$$
$$=6m+3(k+1)(k+2)$$

But if k is odd, $(k+1)$ is even; if k is even, so is $(k+2)$, so $(k+1)(k+2)$ contains one factor that is even, and therefore must be even, and $3(k+1)(k+2)$ is three times an even number, and so has a factor 6. Thus both terms on the right hand side have a factor 6, so $(k+1)(k+2)(k+3)$ is a multiple of 6. But we found that this result was true when $n=2$, so it is true when $n=3$; since it is true when $n=3$, it is true when $n=4$, and so it is true for all positive integral values of n.

Example 4.3 Prove the binomial theorem, that if n is a positive integer

$$(x+y)=x^n+\binom{n}{1}x^{n-1}y+\binom{n}{2}x^{n-2}y^2+\ldots$$

$$\ldots+\binom{n}{r}x^{n-r}y^r+\ldots+x^n$$

where $\quad \binom{n}{r}\equiv{}^nC_r=\dfrac{n!}{r!(n-r)!}$

If this is true for some one value of n, say $n=k$, then

$$(x+y)^k=x^k+\binom{k}{1}x^{k-1}y+\binom{k}{2}x^{k-2}y^2+\ldots+\binom{k}{r}x^{k-r}y^r+\ldots+y^k$$

Multiplying both sides of the equation by $(x+y)$ to obtain $(x+y)^{k+1}$

$$(x+y)^{k+1}=(x+y)(x^k+\binom{k}{1}x^{k-1}y+\binom{k}{2}x^{k-2}y^2+\ldots$$

$$+\binom{k}{r}x^{k-r}y^r+\ldots+y^k)$$

$$=x^{k+1}+\binom{k}{1}x^ky+\binom{k}{2}x^{k-1}y^2+\ldots+\binom{k}{r}x^{k-r+1}y^r+\ldots+xy^k$$

$$+x^ky+\binom{k}{1}x^{k-1}y^2\ldots$$

$$+\binom{k}{r-1}x^{k-(r-1)}y^r+\ldots+y^{k+1}$$

But $\binom{k}{1}+1=k+1=\binom{k+1}{1}$, and more generally

$$\binom{k}{r}\binom{k}{r-1}=\binom{k+1}{r} \text{ (see p86)}$$

$$\therefore \quad (x+y)^{k+1}=x^{k+1}+\binom{k+1}{1}x^ky+\binom{k+1}{2}x^{k+1}y^2+\ldots$$

$$+\binom{k+1}{r}x^{k+1-r}y^r+\ldots+y^{k+1}$$

which is the form we expect.

$\therefore$ If it is true for some one value of n, it is true for the 'next'.

But $(x+y)^2=x^2+2xy+y^2$

$$=x^2+\binom{2}{1}xy+y^2$$

∴ It is true for $n=2$. Since it is true for the value $n=2$ it is true for all positive integral values of n greater than 2.

N.B. Sometimes the case $n=1$ is so trivial that it may not be easy to see that any result being proved for general values of n is true even in that case, as when finding the sum of the first one terms in a series, and as here $(x+y)^1=x+y$.

QUESTIONS

Prove by induction

1 $\sum_1^1 r \equiv 1+2+3+4+\ldots+n=\frac{1}{2}n(n+1)$

2 $\sum_1^n r \equiv 1^2+2^2+3^2+\ldots+n^2=\frac{1}{6}n(n+1)(2n+1)$

3 $\sum_1^n r(r+2) \equiv 1\times3+2\times4+3\times5+\ldots+n(n+2)=\frac{1}{6}n(n+1)(2n+7)$

4 $\sum_{r=0}^n ap^r \equiv a+ap+ap^2+\ldots+ap^n=a\dfrac{p^{n+1}-1}{p-1}$

5 $\sum_{r=1}^n [a+(r-1)d] \equiv a+(a+d)+\ldots+(a+(n-1)d)=\frac{1}{2}n\{2a+(n-1)d\}$

6 That if £P is invested at $r\%$ per annum Compound Interest, the value £A of the investment after n years is £$P(1+r/100)^n$.

7 That the sum of the interior angles of an n-sided polygon is $(2n-4)$ right angles.

8 That all numbers of the form $3^{2n}+7$, where n is a positive integer, are multiples of 8.

SERIES: ARITHMETIC AND GEOMETRIC SERIES, BINOMIAL AND OTHER SERIES

CONTENTS

ARITHMETIC SERIES

$$a+(a+d)+(a+2d)+...+a+(n-1)d$$

with n terms, is an arithmetic series; the sum is $\frac{1}{2}n[2a+(n-1)d]$
If the last term l is known, the sum is $\frac{1}{2}n(a+l)$.

GEOMETRIC SERIES

$$a+ar+ar^2+ar^3+...+ar^{n-1},$$

with n terms, is a geometric series; the sum is $a\,\dfrac{r^n-1}{r-1}$

If the common ratio r is such that $-1<r<1$, the series **converges** and is said to have a **'sum to infinity'** S_∞ where

$$S_\infty = \frac{a}{1-r}$$

BINOMIAL SERIES

$$(x+a)^n = x^n + nx^{n-1}a + \frac{n(n-1)}{1\times2}x^{n-2}a^2 + ...$$

$$... + \frac{n!}{r!(n-r)!}x^{n-r}a^r...+a^n$$

for all values of x and a, when n is a positive integer;

$$(1+x)^n = 1 + nx + \frac{n(n-1)}{1\times2}x^2 + \frac{n(n-1)(n-2)}{1\times2\times3}x^3...$$

for all values of n if $-1<x<1$, and when $x=-1$ and/or $+1$ for some values of n.

LOGARITHMIC SERIES

$$\ln(1+x) = x - \tfrac{1}{2}x^2 + \tfrac{1}{3}x^3 - .. + (-1)^{r-1}\frac{x^r}{r}...$$

when $-1<x\leqslant1$

EXPONENTIAL SERIES

$$e^x = 1 + x + \frac{x^2}{2!} + \frac{x^3}{3!} + \ldots + \frac{x^r}{r!} + \ldots$$

for all values of x.

TRIGONOMETRIC SERIES

$$\sin x = x - \frac{x^3}{3!} + \frac{x^5}{5!} - \frac{x^7}{7!} + \ldots$$

$$\cos x = 1 - \frac{1}{2}x^2 + \frac{x^2}{4!} - \ldots$$

for all values of x.

ARITHMETIC SERIES

The series $a + (a+d) + (a+2d) + \ldots$ in which each term differs from the previous term by the same amount d, is called an arithmetic series (or arithmetic progression); the nth term in the series is $a + (n-1)d$, and the sum of the first n terms $\frac{1}{2}n\{2a + (n-1)d\}$, or $\frac{1}{2}n(a+l)$, if the first and last terms are given but not the common difference.

Example 5.1 The series 1, 5, 9, 13 is an arithmetic series, common difference 4. The twentieth term is $1 + 19 \times 4$, i.e. 77, and the sum of the first twenty terms is

$$\tfrac{1}{2} \times 20 \times \{2 \times 1 + 19 \times 4\}, \text{ i.e. } 780$$

Example 5.2 How many terms in the arithmetic series 2, 3.5, must be taken if the sum is to be 123?

We know that $a = 2$, and can see that $d = 1.5$, so using
$$S = \tfrac{1}{2}n\{2a + (n-1) \times d\}$$
we have $123 = \tfrac{1}{2}n\{4 + (n-1) \times 1.5\}$
$$= \tfrac{1}{2}n\{1.5n + 2.5\}$$
i.e. $3n^2 + 5n - 492 = 0$
$$(3n + 41)(n - 12) = 0$$

$n = 12$ or $-41/3$. But since n is the number of terms, n must be positive, so that $n = 12$, we have to take 12 terms if the sum is to be 123.

GEOMETRIC SERIES

The series a, ar, ar^2, ar^3 … in which each term is in a constant ratio to the previous term is called a geometric series (or geometric progression); the nth term of the series is ar^{n-1} and the sum of the first n terms in the series is

$$a\frac{r^n-1}{r-1} \text{ or } a\frac{1-r^n}{1-r} \text{ if } 0<r<1$$

Example 5.3 The series 2, 1, $\frac{1}{2}$, $\frac{1}{4}$ is a geometric series, common ratio $\frac{1}{2}$. The tenth term is $2\times(\frac{1}{2})^9$, i.e. $(\frac{1}{2})^8$, and the sum of the first ten terms is

$$2\frac{1-(\frac{1}{2})^{10}}{1-\frac{1}{2}}$$

i.e. $4-(\frac{1}{2})^8$. We can see that as the number of terms summed increases, the sum gets closer and closer to 4; this is called the sum to infinity, and the sum to infinity of the geometric series a, ar, ar^2, ... in which

$$r<1, \text{ is } \frac{a}{1-r}.$$

Example 5.4 Show the sum of the first twenty terms of the geometric series 3, 6, 12, ... is greater than 3×10^6, given $2^{10}>1000$

Using the formula $S=a\frac{r^n-1}{r-1}$

$$S=3\frac{2^{20}-1}{2-1}$$

$$=3(2^{20}-1)$$

Since $2^{10}=1024>1000$, $2^{20}>10^6$, so $S>3\times10^6$.

ARITHMETIC MEAN AND GEOMETRIC MEAN

The arithmetic mean of a set of n numbers x_1, x_2, x_3 ... is

$\frac{1}{n}\{x_1+x_2+x_3+...+x_n\}$; the geometric mean of the same set of numbers

is $\sqrt[n]{x_1 \, x_2 \, x_3 \, ... \, x_n}$. For example, the arithmetic mean of 3, 6, 96, is $\frac{1}{3}(3+6+96)$, i.e. 35; the geometric mean is $\sqrt[3]{3\times6\times96}$, i.e., 12.

Example 5.5 Prove that the arithmetic mean of two positive unequal numbers a, b is always greater than their geometric mean.

The arithmetic mean A is $\frac{1}{2}(a+b)$; the geometric mean G is $\sqrt{ab}$. Consider A^2-G^2. Then

$$A^2-G^2=\frac{1}{4}(a+b)^2-ab$$
$$=\frac{1}{4}(a^2+2ab+b^2)-ab$$
$$=\frac{1}{4}(a-b)^2>0 \text{ since } a\neq b$$

Since both A and G are positive,

$$A^2-G^2>0 \Rightarrow A^2>G^2 \Rightarrow A>G$$

CONVERGENT SERIES

We saw, on p49, that the sum of a certain series became 'closer and closer to 4'. The general geometric series a, ar, ar^2 has sum S, where

$$S=a\frac{1-r^n}{1-r}=a\frac{1}{1-r}-a\frac{r^n}{1-r}$$

If $-1<r<1$, $a\frac{r^n}{1-r}$ becomes as small as we wish, and we say that the series has a 'sum to infinity' $a\frac{1}{1-r}$. The series is then called a **convergent** series.

Example 5.6 Find the range of values for which the series

$$1+\frac{x}{3}+\frac{x^2}{9}+\frac{x^3}{27}\ldots$$

has a sum to infinity and an expression in x for that sum.
   The series is a geometric series, common ratio $x/3$, so that there will only be a sum to infinity if $-1<x/3<1$, i.e. $-3<x<3$.
   If x takes a value in that range, the sum to infinity is

$$\frac{1}{1-x/3} \text{ i.e. } \frac{3}{3-x}$$

BINOMIAL SERIES

The binomial theorem states that, if n is a positive integer, for all values of x and y,

$$(x+y)^n=x^n+\binom{n}{1}x^{n-1}y+\binom{n}{2}x^{n-2}y^2+\ldots+\binom{n}{r}x^{n-r}y^r+\ldots+y^n$$

(proved by Induction on p43). It is also true (though the proof is too difficult at this stage) that

$$(1+x)^n=1+nx+\frac{n(n-1)}{1\times2}x^2+\frac{n(n-1)(n-2)}{1\times2\times3}x^3\ldots$$

for all values of n if $-1<x<1$; it can be shown that this is also true if $x=1$ when $n>-1$ and when $x=-1$ when $n>0$.

Example 5.7 Find the range of values for x for which $(1-2x)^{1/2}$ can be expanded as a power series in x, and the first four terms in that series. Since $n=\frac{1}{2}$, $n>0$ so that the expansion is valid when $-1\le-2x\le1$, i.e. $-\frac{1}{2}\le x\le\frac{1}{2}$.

When x has a value in this range;

$$(1-2x)^{1/2}=1+\tfrac{1}{2}(-2x)+\frac{\tfrac{1}{2}(-\tfrac{1}{2})}{1\times2}(-2x)^2+\frac{\tfrac{1}{2}(-\tfrac{1}{2})(-\tfrac{3}{2})}{1\times2\times3}(-2x)^3...$$

$$=1-x-\tfrac{1}{2}x^2-\tfrac{1}{2}x^3$$

Example 5.8 Assuming x to be sufficiently small compared with a for the expansion to be valid, find the first four terms in the expansion in ascending powers of x of

$$\frac{a^3}{(a^2+x^2)^{3/2}}$$

Since the binomial expansion, when n is not an integer, is written in the form $(1+\;)^n$, write $(a^2+x^2)^{3/2}$ as

$$a^3\left(1+\frac{x^2}{a^2}\right)^{3/2}$$

Then

$$\frac{a^3}{a^3\left(1+\dfrac{x^2}{a^2}\right)^{3/2}}=\left(1+\frac{x^2}{a^2}\right)^{-3/2}$$

$$=1+\left(-\frac{3}{2}\right)\!\left(\frac{x^2}{a^2}\right)+\frac{(-\tfrac{3}{2})(-\tfrac{5}{2})}{1\times2}\left(\frac{x^2}{a^2}\right)^2$$

$$+\frac{(-\tfrac{3}{2})(-\tfrac{5}{2})(-\tfrac{7}{2})}{1\times2\times3}\left(\frac{x^2}{a^2}\right)^3$$

$$=1-\frac{3x^2}{2a^2}+\frac{15x^4}{8a^4}-\frac{35x^6}{16a^6}$$

EXPANSION WHEN X IS LARGE

When x is 'large', $1/x$ is 'small', so that it is often possible to expand $(1+1/x)^n$ as a power series in $1/x$.

Example 5.9 If x is sufficiently large for the expansion to be valid, find the first four terms in the expansion of $x/(2+x)^2$ in ascending powers of $1/x$.

$$\frac{x}{(2+x^2)} = \frac{x}{x^2\left(\frac{2}{x^2}+1\right)} = \frac{1}{x}\left(1+\frac{2}{x^2}\right)^{-1}$$

$$= \frac{1}{x}\left[1+(-1)\left(\frac{2}{x^2}\right)+\frac{(-1)(-2)}{1\times2}\left(\frac{2}{x^2}\right)^2\right.$$

$$\left.+\frac{(-1)(-2)(-3)}{1\times2\times3}\left(\frac{2}{x^2}\right)^3 \cdots\right.$$

$$\approx \frac{1}{x}-\frac{2}{x^3}+\frac{4}{x^5}-\frac{8}{x^7}$$

USE OF THE BINOMIAL EXPANSION TO FIND APPROXIMATIONS

Before the ready availability of electronic calculators, the binomial expansion could be used to give approximations to various arithmetic expressions. Now the calculator can be used to check easily an expansion, even if arithmetic approximations are not required.

Example 5.10 Find the first four terms in the binomial expansion of $(1-3x)^{-1/3}$ in ascending powers of x.

Using the binomial theorem,

$$(1-3x)^{-1/3}=1+(-\tfrac{1}{3})(-3x)+\frac{(-\tfrac{1}{3})(-\tfrac{4}{3})}{1\times2}(-3x)^2$$

$$+\frac{(-\tfrac{1}{3})(-\tfrac{4}{3})(-\tfrac{7}{3})(-3x)^3}{1\times2\times3} \cdots$$

$$=1+x+2x^2+\tfrac{14}{3}x^3\ldots$$

Putting $x=0.01$,

$$(1-0.03)^{-1/3}=(0.97)^{-1/3}$$

$$=1.010\ 204\ 7\ldots$$

Substituting in the expansion,

$$=1+0.01+2(0.01)^2+\tfrac{14}{3}(0.01)^3$$
$$=1.010\ 204\ 69\ldots$$

so our expansion looks likely to be correct.

LOGARITHMIC SERIES

It can be proved by Maclaurin's theorem (p272) that if $-1 < x \leqslant 1$;

$$\ln(1+x) = x - \tfrac{1}{2}x^2 + \frac{x^3}{3} \ldots (-1)^{r-1}\frac{x^r}{r} \ldots$$

(Check by substituting $x = -0.01$ in $\ln(1+x)$ and in the expansion.)

Example 5.11 Find the expansion of $\ln\left(\frac{(1+x^2)^{1/2}}{1-2x}\right)$ in ascending powers of x, up to and including the term in x^4, assuming that the value of x is such that the expansion is valid.

$$\ln\frac{(1+x^2)^{1/2}}{1-2x} = \tfrac{1}{2}\ln(1+x^2) - \ln(1-2x)$$

$$= \tfrac{1}{2}\left[x^2 - \frac{x^4}{2} \ldots\right]$$

$$- [-2x - \tfrac{1}{2}(-2x)^2 + \tfrac{1}{3}(-2x)^3 - \tfrac{1}{4}(-2x)^4 \ldots]$$

$$= \frac{x^2}{2} - \frac{x^4}{4} \ldots - [-2x - 2x^2 - \tfrac{8}{3}x^3 - 4x^4 \ldots]$$

$$= 2x + \tfrac{5}{2}x^2 + \tfrac{8}{3}x^3 + \tfrac{15}{4}x^4 \ldots$$

For the expansion to be valid, $-1 < y \leqslant 1$, i.e. $-1 < x^2 \leqslant 1$ and also $+1 < -2x \leqslant 1$. Both these conditions are satisfied only if $-\tfrac{1}{2} \leqslant x < \tfrac{1}{2}$.

EXPONENTIAL AND TRIGONOMETRIC SERIES

We can also use Maclaurin's theorem to prove that

$$e^x = 1 + x + \frac{x^2}{2!} + \frac{x^3}{3!} \cdots \frac{x^r}{r!} \cdots$$

$$\cos x = 1 - \frac{x^2}{2!} + \frac{x^4}{4!} \cdots$$

and $\quad \sin x = x - \frac{x^3}{3!} + \frac{x^5}{5!}$

These expansions are valid for all (real) values of x.

Example 5.12 Find the coefficient of x^n in the expansion of e^{1+2x} in ascending powers of x.

$$e^{1+2x} = e(e^{2x})$$

$$= e\left(1 + 2x + \frac{(2x)^2}{2!} + \frac{(2x)^3}{3!} \dots\right)$$

$$= e + 2ex + \frac{2^2ex^2}{2!} + \frac{2^3ex^3}{3!} \dots \frac{2^nex^n}{n!}$$

thus the coefficient of x^n is $\dfrac{2^ne}{n!}$

More questions and examples on the logarithmic, exponential and trigonometric series are gives in Chapter 21.

QUESTIONS

1   Which of the following series is an arithmetic series, which a geometric series, and which neither?

   (a)   1, 2.5, 5, ...
   (b)   1, 2, 3, 5, ...
   (c)   3, 18, 108, ...
   (d)   0.6, 1.2, 1.8, ...
   (e)   −0.8, 0.8, 1.8, ...

2   If the rth term of a series is denoted by u_r, in which of the following cases is the series arithmetic and which geometric?

   $u_r =$ (a) $2r$ (b) $2-r$ (c) 2^r (d) 2^{-r}
          (e) $1 - \frac{1}{2}r$

3   The third term of an arithmetic series is 3 and the sixth term is 12. Find the first term and the sum of the first ten terms.

4   The fifth term of an arithmetic series is 8 and the ninth term is 14. Find the fourth term and the sum of the first twenty terms.

5   Find the first negative term in the arithmetic series
     8, 6.5, 5, ...

6   The fourth term of a geometric series is 3 and the sixth term is $\frac{3}{16}$. Find the two possible values of the common ratio and of the first term.

7   How many terms of the arithmetic series 1, 5, 9, ... must be taken if the sum is 190?
     What is the least number of terms of this series that must be taken if the sum is to exceed 900?

8   Find the term in the arithmetic series 2, 7, 12 which is greater than 1000.

9   Find the term in the geometric series 2, 10, 50 which is greater than 1000.

10   What is the least number of terms in the geometric series 2, 6, 18, ... if the sum exceeds 1000?

11 Expand each of the following, giving the expansions in their simplest form.

   (a) $(2+x)^3$ (b) $(1+2x)^3$ (c) $(2-3x)^3$
   (d) $(1-3x)^4$ (e) $(2-3x)^4$ (f) $(2+3x)^4$

12 Find the first four terms when each of the following expressions is expanded in ascending powers of x, and the range of values for which each expansion is valid.

   (a) $(1-2x)^{-1}$ (b) $(1-2x)^{-2}$
   (c) $(1-2x)^{-\frac{1}{2}}$ (d) $(1-2x)^{\frac{1}{2}}$
   (e) $(1-2x^2)^{-1}$ (f) $(2-x)^{-1}$

13 Find the first three terms when each of the following is expanded in ascending powers of x and state the range of values of x for which the expansion is valid.

   (a) $(1+x^2)^{-1}$ (b) $(1+x^2)^{-2}$ (c) $(1+x^2)^{-3}$
   (d) $(1-2x^2)^{-1}$ (e) $x(1-3x^2)^{-1}$ (f) $x^2(1-x^2)^{-1}$

14 Find the first four terms when each of the following expressions is expanded in ascending powers of $1/x$.

   (a) $(x-1)^{-1}$ (b) $(x^2+2)^{-1}$
   (c) $(x+1)^{-2}$ (d) $(x^2-2)^{-2}$

15 Find the expansions in ascending powers of x of $(1-4x)^{\frac{1}{2}}$ and $1-2x(1-ax)^b$, up to and including the term containing x^3. Find the values of a and b if these expansions are identical for the terms that have been found.

16 Write down the first four non-zero terms in each of the following assuming the values of x to be such that the expansion is valid:

   (a) $\ln(1-x)$ (b) $\ln(1-2x)$
   (c) $\ln(1+3x)$ (d) $\ln(1+3x^2)$
   (e) $\ln(1-4x^2)$ (f) e^{2x}
   (g) e^{x+1} (h) e^{x^2+1}
   (i) $\sin(2x)$ (j) $\cos(x^2)$

17 Factorise $1-3x+2x^2$ and hence find the first four terms in the expansion of $\ln(1-3x+2x^2)$ as a power series in x. Check by finding the value of $\ln(0.72)$.

18 Find the first four terms in the expansion of $\ln\left(\dfrac{1+3x}{1-2x}\right)$ and give a numerical check for your expansion.

19 Write down the first four non-zero terms in the expansions of $\sin x$ and $\cos x$, obtain the product $\sin x \cos x$ up to and including the term in x^8 and compare with the expansion of $\sin 2x$.

20 Find the first four non-zero terms in the expansions of

   (a) $\frac{1}{2}(e^x+e^{-x})$ (b) $\frac{1}{2}(e^x-e^{-x})$

21 The rth term of an arithmetic series is $3r-1$. Find the first term and the sum of the first 20 terms.

22 Find x if 1, x, 7, 10 are the first four terms of an arithmetic series, and find how many terms of this series must be taken for the sum to exceed 1000.

23　The rth term of a geometric series is $8(\frac{1}{2})^r$. Find the 10th term of the series, and the sum to infinity of the series.

24　Find the term that does not contain x in the expansion of

$$\left(x^2-\frac{1}{x}\right)^4\left(x+\frac{1}{x}\right)^3$$

25　Given $a<b$, find a and b if the first three terms in the expansion of

$$\frac{1+ax}{\sqrt{1+bx}}$$

in ascending powers of x are $1+2x-\frac{3}{2}x^2$

26　Express $2(1-2x)^{\frac{1}{2}}-(4+5x)^{-\frac{1}{2}}$ as a power series in x, up to and including the term in x^2.

27　Given that x is sufficiently small, find the constants a, b and c if

$$\left(\frac{1+3x}{8+3x}\right)^{\frac{1}{3}}\approx a+bx+cx^2$$

28　Show that the first three terms in the expansion of $(1-8x)^{\frac{1}{4}}$ in ascending powers of x are the same as the first three terms in the expansion

of $\dfrac{1-5x}{1-3x}$

29　Find a and b if the first three terms in the expansion of

$$\frac{(1+ax)^4}{(1+bx)^3} \text{ are } 1-x+6x^2, \text{ given } a \text{ and } b \text{ are both positive.}$$

30　Find correct to the nearest whole number, the sum of the first 20 terms of the geometric series $1+e^{1/2}+e+e^{3/2}...$; find also the range of values of x for which the geometric series $1+e^x+e^{2x}...$ has a sum to infinity S, and find the value of S, correct to four significant figures when $x=-0.5$.

31　Find the values of a and b if the first three terms in the expansion of $e^{ax}/(1-bx^2)$ are $1+\frac{1}{2}x+\frac{3}{8}x^2$.

32　Write down the first three terms in the expansion of $u=\left(1+\dfrac{x}{n}\right)^n$ in ascending powers of x. Show that if all terms in x^3 and higher powers of x are neglected, then

$$u+\frac{1}{u}=2e^x$$

33　Write down and simplify the expansions of $\ln(2-x)$ and $\ln[(2-x)(1-x)]$, up to and including the terms in x^3. Put $x=0.2$, to find an approximation for $\ln(1.2)$, and check your approximation using a calculator.

34　Find the values of a and b if $e^x \ln(1+x)=ax+bx^2$, when x is sufficiently small for terms in x^3 to be neglected.

35　Find the values of a and b if $(x+a \sin x)(1-\cos x)=bx^5$, when terms in x^6 and higher powers are neglected.

PARTIAL FRACTIONS

CONTENTS

NOTES

EQUATIONS AND IDENTITIES

An equation is only satisfied by **some** values of x; an identity is true for **all** values of x.

PARTIAL FRACTIONS

Remember that the degree of the numerator of the fraction must be less than the degree of the denominator; if it is not less, we must first divide, e.g.

$$\frac{x^2}{(x-1)(x+1)} = \frac{x^2-1+1}{(x-1)(x+1)} = 1 + \frac{1}{(x-1)(x+1)}$$

$$= 1 + \frac{\frac{1}{2}}{x-1} - \frac{\frac{1}{2}}{x+1}$$

Suitable partial fractions are:

$$\frac{c}{(x-a)(x-b)} = \frac{A}{(x-a)} + \frac{B}{(x-b)}$$

$$\frac{c}{(x-a)(x^2+b)} = \frac{A}{(x-a)} + \frac{Bx+C}{(x^2+b)}$$

$$\frac{c}{(x-a)(x-b)^2} = \frac{A}{(x-a)} + \frac{B}{(x-b)} + \frac{C}{(x-b)^2}$$

EQUATIONS AND IDENTITIES

$x^2-1=0$ is true for only two values of x, $x=1$ or -1;

$x^2-1=x^2$ is not true for any value of x;

$x^2-1=(x-1)(x+1)$ is true for all values of x and is called an *identity*. It is often written $\equiv$.

Quadratic equations have at most two distinct real roots, and polynomials in x of degree n will be satisfied by at most n different values of x. If an equation in x is satisfied by *more* than n different values of x, then it is an identity and is satisfied by all values of x. Thus two expressions are identically equal if the coefficients of each term are equal, or if they have the same values for all values of x.

Example 6.1 Show that $x(x-1)-2(x-1)(x-2)+x(x-2)=3x-4$ is true for all values of x.

Method 1: Expanding the brackets on the L.H.S.,

$$x(x-1)-2(x-1)(x-2)+x(x-2)$$
$$=x^2-x-2x^2+6x-4+x^2-2x$$
$$=3x-4,\text{ the same expression as the R.H.S.}$$

Method 2: This is an equation of degree two, so it will be an identity if it is satisfied by *more than two values* of x.

Put $x=0$;   L.H.S.$=-4$,   R.H.S.$=-4$
Put $x=1$;   L.H.S.$=-1$,   R.H.S.$=-1$
Put $x=2$;   L.H.S.$=2$ R.H.S.$=2$

Thus the equation is satisfied by at least three values of x and so is an identity.

N.B. The values $x=0$, 1 and 2 were chosen so that as many as possible of the brackets on the L.H.S. were zero.

Example 6.2 Show that $x(x-1)-2(x-1)(x-2)+x(x-2)=3x-5$ is not satisfied by any values for x.

Expanding the L.H.S. as before,

$$\text{L.H.S.}=3x-4$$

so the equation becomes

$$3x-4=3x-5$$

which is not true, whatever the value of x.

PARTIAL FRACTIONS

We can see that $$\frac{1}{x}+\frac{1}{x+1}\equiv\frac{2x+1}{x(x+1)}$$

and that $$\frac{1}{x}-\frac{1}{x+1}\equiv\frac{1}{x(x+1)}$$

Can we find constants A and B so that, say

$$\frac{2x-1}{x(x+1)}\equiv\frac{A}{x}+\frac{B}{x+1}?$$

If this is so, multiply both sides by $x(x+1)$

$$2x-1\equiv A(x+1)+Bx$$

Put $x=0$, then $A=-1$; put $x=-1$, then $B=3$, so that

$$2x-1\equiv-(x+1)+3x$$

i.e. $\quad \dfrac{2x-1}{x(x+1)} = \dfrac{-1}{x} + \dfrac{3}{x+1}$

There are three types of fractions that we have to express as partial fractions at this stage,

(1) where the factors of the denominator are linear and all different, e.g.

$$\frac{1}{(x-2)(x+3)}$$

(2) where one of the factors in the denominator is quadratic, e.g.

$$\frac{1}{(x-2)(x^2+3)}$$

(3) where one of the factors in the denominator is repeated, e.g.

$$\frac{1}{(x-2)(x+3)^2}$$

In each case, before we can express a 'fraction' in partial fractions we have to ensure that it is a fraction, i.e. that the degree of the numerator is less than the degree of the denominator. (If this is not already so, then we have to divide – see Example 6.7.) We can then assume that the degree of the numerator of every fraction is at least one less then the degree of its denominator, e.g.

$$\frac{A}{x-2} \text{ and } \frac{Bx+C}{x^2+3} \text{ type (2)}$$

occasionally B or C may be zero. In type (3), however, we need only have constant terms in each denominator, so that the partial fractions will be of the form

$$\frac{A}{(x-2)} + \frac{B}{(x+3)} + \frac{C}{(x+3)^2}$$

Example 6.3 Find the values of A and B if

$$\frac{5}{(x-2)(x+3)} \equiv \frac{A}{(x-2)} + \frac{B}{(x+3)}$$

Method 1: Multiply both sides by $(x-2)(x+3)$, and find the values for A and B if

$$5 = A(x+3) + B(x-2) \qquad (1)$$

is true for all values of x

i.e. $\qquad\qquad\qquad 5 \equiv Ax + 3A + Bx - 2B$

Equating coefficients of x $\qquad 0 = A + B$

Equating constants, $\qquad\qquad 5 = 3A - 2B$

Solving, $A=1$ and $B=-1$

$\therefore$
$$\frac{5}{(x-2)(x+3)}=\frac{1}{x-2}-\frac{1}{x+3}$$

Method 2: In (1), put $x=-3$,

then $\qquad$ $5=B(-5)$, i.e. $B=-1$

Put $x=2$, then $\qquad$ $5=A(5)$, i.e. $A=1$ $\qquad\qquad$ (2)

Method 3: 'Cover up' or 'finger' method.
Dividing equation (1) by $(x+3)$ we have

$$\frac{5}{(x+3)}=A+\frac{B(x-2)}{(x+3)}$$

When $x=2$,

$$\frac{5}{(2+3)}=A, \text{ as in (2), giving } A=1.$$

We see that the value of A is that obtained by 'covering up' the factor $(x-2)$ in $\frac{5}{(x-2)(x+3)}$ and substituting $x=2$ in what remains to get

Fig 6.1

and the value of B is that obtained by covering up the factor $(x+3)$ in $\frac{5}{(x-2)(x+3)}$ and substituting $x=-3$ in what remains, i.e.

Fig 6.2

i.e. $B=-1$.

Since we often 'cover up' with a finger this is sometimes called the 'finger' method.

Example 6.4 Find the values of A, B and C if

$$\frac{12}{(x+1)(x-2)(x-3)}=\frac{A}{(x+1)}+\frac{B}{(x-2)}+\frac{C}{(x-3)}$$

Cover up the factor $(x+1)$ and put $x=-1$ in what remains

$$\frac{12}{(-3)(-4)}=A, \text{ i.e. } A=1$$

Cover up the factor $(x-2)$ and put $x=2$ in what remains

$$\frac{12}{(3)\qquad(-1)}=B, \text{ i.e. } B=-4$$

Cover up the factor $(x-3)$ and put $x=3$ in what remains

$$\frac{12}{(4)(1)}=C \text{ i.e. } C=3$$

$$\frac{12}{(x+1)(x-2)(x-3)}=\frac{1}{(x+1)}-\frac{4}{(x-2)}+\frac{3}{(x-3)}$$

Check: Multiply both sides by $(x+1)((x-2)(x-3)$

$$12=1(x-2)(x-3)-4(x+1)(x-3)+3(x+1)(x-2)$$

The coefficient of x^2 on the L.H.S. is 0; on the R.H.S. is $1-4+3=0$

Example 6.5 Find the values of A, B and C if

$$\frac{3}{(x+1)(x^2+2)}\equiv\frac{A}{(x+1)}+\frac{Bx+C}{(x^2+2)}$$

Notice that we have to take the numerator of the second term as $Bx+C$ since the degree of the denominator is two.

The value of A can be found by the 'cover up' method, but we shall here use a combination of methods 1 and 2.

Multiply both sides by $(x+1)(x^2+2)$

$$3=A(x^2+2)+(x+1)(Bx+C)$$

Put $x=-1$, then $\qquad 3=A[(-1)^2+2]$, i.e. $A=1$

Equating the coefficients of x^2,

$$0=A+B, \quad \therefore B=-1$$

Equating the constants, $3=2A+C$, $\therefore C=1$

Check: equate the coefficients of x. On the L.H.S., 0; on the R.H.S. $B+C=-1+1=0$.

$$\therefore \qquad \frac{3}{(x+1)(x^2+2)}\equiv\frac{1}{(x+1)}+\frac{-x+1}{(x^2+2)}$$

Example 6.6 Find the values of A and B if

$$\frac{3x+7}{(x+1)^2} \equiv \frac{A}{(x+1)} + \frac{B}{(x+1^2)}$$

Method 1: Multiply both sides by $(x+1)^2$

$$3x+7 \equiv A(x+1)+B$$

Put $x = -1$, $4 = B$

Equating the coefficients of x, $3 = A$

$\therefore$

$$\frac{3x+7}{(x+1)^2} \equiv \frac{3}{(x+1)} + \frac{4}{(x+1)^2}$$

Method 2: Substitute $x+1 = y$

Then $\dfrac{3x+7}{(x+1)^2} \equiv \dfrac{3y+4}{y^2} \equiv \dfrac{3}{y} + \dfrac{4}{y^2} \equiv \dfrac{3}{(x+1)} + \dfrac{4}{(x+1)^2}$

With a little practice this can be worked through without actually making the substitution, i.e.

$$\frac{3x+7}{(3x+1)^2} \equiv \frac{3(x+1)+4}{(x+1)^2} \equiv \frac{3}{x+1} + \frac{4}{(x+1)^2}$$

Example 6.7 Express in partial fractions

$$\frac{x^4}{(x-1)(x^2+2)}$$

In this case the degree of the numerator is 4, of the denominator is 3 so we must first divide. We can use long division, or the identity $x^4 \equiv (x+1)(x^3-x^2+2x-2)-x^2+2$, since $(x-1)(x^2+2) \equiv x^3-x^2+2x-2$.

Then $\dfrac{x^4}{(x-1)(x^2+2)} \equiv \dfrac{(x+1)(x^3-x^2+2x-2)-x^2+2}{(x-1)(x^2+2)}$

$$\equiv x+1 - \frac{x^2-2}{(x-1)(x^2+2)}$$

Now if $\dfrac{x^2-2}{(x-1)(x^2+2)} \equiv \dfrac{A}{x-1} + \dfrac{Bx+C}{x^2+2}$

$$x^2-2 \equiv A(x^2+2)+(Bx+C)(x-1)$$

Put $x=1$, then $A = -1/3$
Equating the coefficients of x^2 $1 = A+B \therefore B = \frac{4}{3}$
Equating the constants $-2 = 2A-C \therefore C = \frac{4}{3}$

To check, equate the coefficients of x, $0=-B+C$, which is so.

$$\frac{x^4}{(x-1)(x^2+2)}=x+1+\frac{\frac{1}{3}}{x-1}-\frac{\frac{4}{3}(x+1)}{x^2+2}$$

APPLICATIONS OF PARTIAL FRACTIONS

Partial fractions are often useful when finding binomial expansions or when integrating.

Example 6.8 Express in partial fractions and find the first two non-zero terms in the expansion as a power series in x of

$$\frac{x}{(x-2)(x-3)}$$

Use the 'cover up' method and put $x=2$;

$$A=\frac{2}{(2-3)}=-2$$

Put $x=3$,

$$B=\frac{3}{(3-2)}=3$$

$\therefore$

$$\frac{x}{(x-2)(x-3)}=\frac{-2}{x-2}+\frac{3}{x-3}$$

Remember that the denominators must be written in the form $(1-\quad)$ before the binomial expansion is used.

$$\text{R.H.S.}=\frac{1}{1-\frac{x}{2}}-\frac{1}{1-\frac{x}{3}}$$

$$=\left(1-\frac{x}{2}\right)^{-1}-\left(1-\frac{x}{3}\right)^{-1}$$

$$=1+(-1)\left(-\frac{x}{2}\right)+\frac{(-1)(-2)}{1\times2}\left(-\frac{x}{2}\right)^2\cdots$$

$$-1-(-1)\left(-\frac{x}{3}\right)-\frac{(-1)(-2)}{1\times2}\left(-\frac{x}{3}\right)^2\cdots$$

$$=\frac{x}{6}+\frac{5x^2}{36}, \text{ neglecting higher powers of } x.$$

Example 6.9 Find

$$\int \frac{1}{(x-1)(x-2)}dx$$

Since $\dfrac{1}{(x-1)(x-2)}=\dfrac{-1}{x-1}+\dfrac{1}{x-2}$

using the 'cover up' or any other method,

$$\int \frac{1}{(x-1)(x-2)}dx=\int \left(\frac{-1}{x-1}+\frac{1}{x-2}\right)dx$$

$$=-\ln|x-1|+\ln|x-2|+C$$

$$=\ln\left|\frac{x-2}{x-1}\right|+C$$

QUESTIONS

1. Show that the following are true for all values of x:
   (a) $x(x+1)-2(x+1)(x+2)+x(x+2)=-3x-4$
   (b) $x(x-1)-3(x-1)(x-2)+2x(x-2)=4x-6$
2. Show that the following are not true for any value of x:
   (a) $x(x-1)+2x(x-3)=3x^2-7x+1,$
   (b) $(x(x+2)+(x-3)(x+3)=2x^2+2x$
3. Show that each of the following is true for some values of x, and find those values:
   (a) $(x-1)(x+1)=8$ (b) $(x-1)(x+2)=4$
   (c) $(x-1)(x+2)=x^2+4$ (d) $(x-1)(x+2)=x^2-2$
4. Which of the following are identities?
   (a) $x^3=(x+1)(x^2-x+1)$
   (b) $x^3+1=(x+1)(x^2-x+1)$
   (c) $x^3+x=(x+1)(x^2-x+1)$
5. Find the constants A, B and C as appropriate in the following:

   (a) $\dfrac{1}{(x+1)(x+2)}\equiv\dfrac{A}{x+1}+\dfrac{B}{x+2}$

   (b) $\dfrac{x}{(x+1)(x+2)}\equiv\dfrac{A}{x+1}+\dfrac{B}{x+2}$

   (c) $\dfrac{x^2}{(x+1)(x+2)}\equiv1+\dfrac{A}{x+1}+\dfrac{B}{x+2}$

   (d) $\dfrac{1}{(x+1)(x^2+2)}\equiv\dfrac{A}{x+1}+\dfrac{Bx+C}{x^2+2}$

(e) $\quad \dfrac{x}{(x+1)^2} \equiv \dfrac{A}{x+1} + \dfrac{B}{(x+1)^2}$

(f) $\quad \dfrac{x}{(x+1)(x+2)(x+3)} \equiv \dfrac{A}{x+1} + \dfrac{B}{x+2} + \dfrac{C}{x+3}$

(g) $\quad \dfrac{1}{x(x+1)^2} \equiv \dfrac{A}{x} + \dfrac{B}{x+1} + \dfrac{C}{(x+1)^2}$

(h) $\quad \dfrac{x^3}{(x-1)(x+1)} \equiv x + \dfrac{A}{x-1} + \dfrac{B}{x+1}$

6 Express in partial fractions:

(a) $\quad \dfrac{2x+3}{(x-2)(x-3)}$ (b) $\quad \dfrac{2x+3}{(x-2)(x^2+3)}$

(c) $\quad \dfrac{2x+3}{(x-2)(x-3)^2}$ (d) $\quad \dfrac{2x+3}{(x-2)(x^2-9)}$

(e) $\quad \dfrac{2x^2+3}{(x-2)(x-3)}$ (f) $\quad \dfrac{2x^3}{(x-2)(x-3)}$

7 Find the binomial expansion of each of the following, up to and including the term in x^2:

(a) $\quad \dfrac{1}{(2x-1)(x-1)}$ (b) $\quad \dfrac{x}{(2x-1)(x-1)}$

(c) $\quad \dfrac{1}{(2x-1)(x^2+1)}$ (d) $\quad \dfrac{1}{(2x-1)(x+1)^2}$

8 Find the following integrals:

(a) $\quad \displaystyle\int \dfrac{1}{(2x-1)(2x+1)} \, dx$ (b) $\quad \displaystyle\int \dfrac{1}{(x(x^2+1)} \, dx$

(c) $\quad \displaystyle\int \dfrac{1}{(x(x^2-1)} \, dx$ (d) $\quad \displaystyle\int \dfrac{x}{(x-1)^2} \, dx$

INEQUALITIES AND SIMPLE CURVE SKETCHING

CONTENTS

INEQUALITIES

If $\quad ax=b$ and $a \neq 0, x = b/a$

but if $\quad ax < b, x < b/a$ **only if a is positive.**

If $\quad (x-a)(x-b)=0, x=a$ or b

but if $\quad (x-a)(x-b)<0, a<x<b$

and if $\quad (x-a)(x-b)>0, x, <a$ or $x>b$

CURVE SKETCHING

Become familiar with as many sketches as possible, and notice the relation between the sketch-graphs of similar functions, e.g.

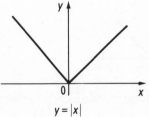

$y = |x|$ $\qquad$ *and*

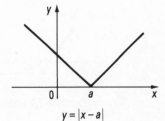

$y = |x-a|$

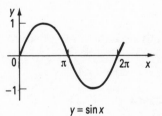

$y = \sin x$ $\qquad$ *and*

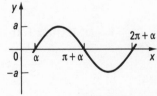

$y = a \sin(x - \alpha)$

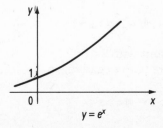

$y = e^x$ $\qquad$ *and*

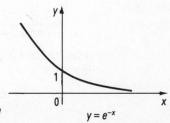

$y = e^{-x}$

Fig 7.1

INVESTIGATE ...

1 the domain over which the function is defined, and the range of values attained ...
2 any symmetry ...
3 where the curve crosses the coordinate axes ...
4 whether there are any asymptotes parallel to either coordinate axis ...
5 the shape of the curve when x is small, to find the tangent(s) at the origin.

INEQUALITIES

If $ax=b$ and $a \neq 0$, $x=b/a$, but if $ax<b$, $x<b/a$ only if a is positive. Care must always be taken , when multiplying and dividing by a number, that the number is positive; if the number is negative, the nature of the inequality is altered, e.g. $-3x<6 \Leftrightarrow x>-2$

If $\quad (x-a)(x-b)=0$, then $x=a$ or b

but if $\quad (x-a)(x-b)<0$, then $a<x<b$

and $\quad (x-a)(x-b)>0$, then $x<a$ or $x>b$, if $a<b$

It is often helpful to sketch the graph to find the range of values that satisfy an inequality.

$(x-a)(x-b)>0$ if
$x<a$ or $x>b$, if a is less than b

Fig 7.2

Example 7.1 Find the range of values of x for which

$$\frac{1}{x-2}>\frac{2}{x}$$

Method 1: If $x>2$, both $x-2$ and x are positive,

$$\frac{1}{x-2}>\frac{2}{x} \Rightarrow x>2(x-2) \Rightarrow x<4$$

$\therefore \qquad\qquad 2<x<4$

If $0<x<2$, $x-2$ is negative and

$$\frac{1}{x-2}>\frac{2}{x} \Rightarrow x<2(x-2) \Leftrightarrow x>4$$

which is not consistent with $0<x<2$
If $x<0$, x and $x-2$ are both negative, so

$$\frac{1}{x-2}>\frac{2}{x} \Rightarrow x>2x-4 \Leftrightarrow x<4$$

which is so since x is already <0, $\therefore x<0$ or $2<x<4$ satisfy the original inequality.

Method 2: sketching the curves $y=\dfrac{1}{x-2}$ and $y=2/x$, serves as a check.

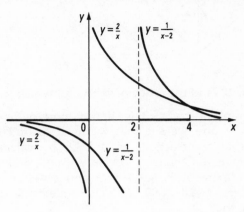

Fig 7.3

From the graphs we can see that $\dfrac{1}{x-2}$ is always greater than $2/x$ when x is negative, and also that it is greater when $2<x<4$. The value $x=4$ has to be found by solving the equation $\dfrac{1}{x-2}=\dfrac{2}{x}$, i.e. $x=2(x-2)$, $x=4$.

$y=|x|$

The expression $|x|=x$ when x is positive or zero, $-x$ when x is negative. It is read 'mod x', and is a special case of $|z|$, the modulus of a complex number. The graph of $y=|x|$ is given in Fig 7.4.

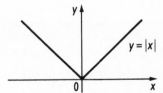

Fig 7.4

Example 7.2 Find the range of values for which $|x|>|x-2|$

We could do this in three stages, considering the ranges of values $x\leqslant 0$, $0\leqslant x\leqslant 2$ and $x\geqslant 2$, but it is easier if we draw the graphs of $y=|x|$ and $y=|x-2|$. We can see from Fig 7.5 that $|x|>|x-2|$ if and only if $x>1$.

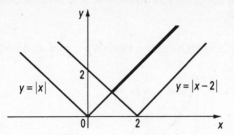

Fig 7.5

Example 7.3 Find the range of values for which $|x|<|2x-3|$

The graph of $y=|2x-3|$ is slightly more difficult to draw than that of $y=|x-2|$, and we see that we need to find the coordinates of the points of intersection, P and Q, by calculation. To find the x coordinate of Q, solve $y=x$ and $y=2x-3$, and obtain $x=3$. To find the x coordinate of P, solve $y=x$ and $y=3-2x$. (Since $x<3/2$, $2x-3$ is negative, so $|2x-3|=3-2x$.) These meet where $x=1$. From the graph we see that

$$|x|<|2x-3| \text{ if } x<1 \text{ or if } x>3.$$

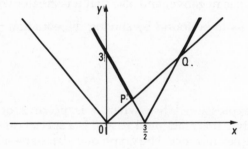

Fig 7.6

CURVE SKETCHING

It is vital to be familiar with some common curves; in the examples above, the graph of $y=\dfrac{1}{x-2}$ was deduced from that of $y=\dfrac{1}{x}$ and the graph of the straight line $y=2x-3$ was deduced from the straight line through the origin $y=x$.

POLYNOMIALS OF THE FORM
$y=a(x-b)(x-c)(x-d)$

This graph crosses the x-axis only where $x=b$, c or d; if a is positive, for large x (i.e. $x>d$), y is positive and the graph has the form shown in Fig 7.7(a). If a is negative, the form of the graph is that in Fig 7.7(b), the reflection of Fig 7.7(a) in the x-axis.

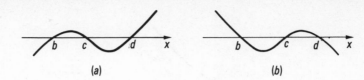

Fig 7.7

(a)   (b)

SOME COMMON CURVES Figures 7.8–7.13 show some common curves with which we must be familiar.

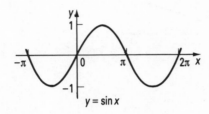

Fig 7.8

$y = \sin x$

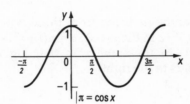

Fig 7.9

$\pi = \cos x$

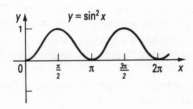

$y = \sin^2 x$

Fig 7.10

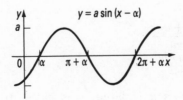

$y = a \sin (x - \alpha)$

Fig 7.11

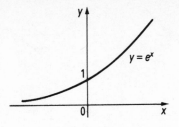

Fig 7.12

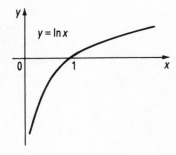

Fig 7.13

Other curves can be deduced from these. The graph of $y=e^{-x}$ is the reflection of Fig 7.12 in the y-axis; $y=\ln(1+x)$ is Fig 7.13 translated one unit to the left, in the same way that Fig 7.11 is Fig 7.8 after a translation α units to the right.

THE GRAPH OF
$y=(x-b)/(x-a)$

The method adopted in sketching this curve illustrates the procedure to be followed in cases where the curve cannot be sketched immediately by comparison with a known curve.

1 Domain and range. Are there any restrictions on the values of x or y? In this case, $x \neq a$.
2 Symmetry. It is not easy to spot any symmetry.
3 Crossing the coordinate axes. When $y=0$, $x=b$; when $x=0$, $y=b/a$.
4 Asymptotes. When $x=a$, y is infinite. Writing the equation of the curve as

$$y = \frac{x-a+a-b}{x-a}$$

i.e. $$y = 1 + \frac{a-b}{x-a}$$

we see that when x is large, y is close to 1, that is, $y=1$ is an asymptote.

5 Shape for small values of x and y. We have found where the graph crosses the axes, and cannot get any more information in this case.

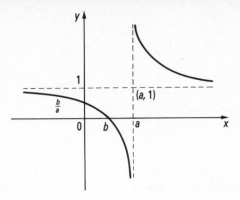

Fig 7.14

To draw the curve, let us suppose that a and b are both positive, and a greater than b. Then we can easily sketch in the asymptote $x=a$, and the points at which the curve crosses the axes. Since $a>b$, when x is large and positive, y is little more than 1; when x is large and negative, y is a little less than 1. This enables us to complete the sketch, as in Fig 7.14.

Having sketched the curve, we see that it has rotational symmetry about $(a, 1)$, but we could hardly have anticipated that from the equation, unless possibly we had written it in the form

$$y-1 = \frac{a-b}{x-a}$$

THE CURVE
$y=ax/(x^2+a^2)$

Follow the same procedure:

1   Domain and range? Clearly the domain can be the set of all real numbers, and there are no obvious restrictions on the range.

2   Symmetry? The function $f:x \rightarrow ax(x^2+a^2)$ is an odd function, so that the curve is symmetrical about the origin. Alternatively, we can say that using the x, y notation, when $x \rightarrow -x$, $y \rightarrow -y$ so the curve has rotational symmetry order two about the origin.

3   Crossing the coordinate axes? When $x=0$, $y=0$, and this is the only point at which it crosses the axes.

4   Asymptotes? There is no value of x for which y is infinite, so there are no asymptotes parallel to the y-axis. When x is large, y is small, so $y=0$ is an asymptote.

5   Shape for small values of x? When x is small compared with a, $y \approx ax/a^2 = x/a$, so that $y=x/a$ is the tangent at the origin. We can now sketch the curve, as in Fig 7.15.

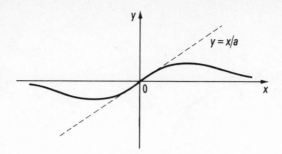

Fig 7.15

Having sketched the curve, we see that there are restrictions on the range, but we could not easily have spotted them, and they did not affect our ability to sketch the curve.

CURVES DESCRIBED BY PRODUCTS OF FUNCTIONS

If the equation of a curve is the product of functions, e.g. $y=xe^{-x}$, it is often helpful to sketch the graph of each function, here $y=x$ and $y=e^{-x}$, and deduce the required sketch from these, as in Fig 7.16.

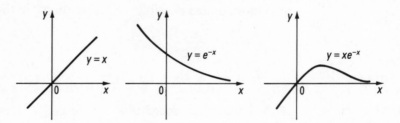

Fig 7.16

QUESTIONS

1   Find the range of values of x for which

(a) $3(x+1)<x-1$ (b) $x-1\geqslant3x-9$

(c) $x^2-5x+6>0$ (d) $\dfrac{5}{x-1}\geqslant\dfrac{1}{x+1}$

(e) $x-5>\dfrac{6}{x}$

2   Given that x is positive, find the range of values of x if

(a) $2(x+1)<x+5$ (b) $x^2-4x+3\geqslant0$

(c) $x-2>\dfrac{3}{x}$ (d) $\dfrac{2}{x+4}<\dfrac{1}{x+1}$

3   Given that x is greater than 2, find the range of values of x if

(a) $2(x+1)<x+5$ (b) $x^2-4x+3\geqslant0$

4 Find the range of values of x for which

(a) $|x|>2$ (b) $|x|>2x$ (c) $|x|>6-2x$

(d) $|3-2x|>|x+5|$ (e) $|x-2|\leqslant|2x+1|$

5 Sketch the following curves, marking the points at which each meets the coordinate axes:

(a) $y=(x-2)(x+4)$ (b) $y=x(x-1)(x+3)$
(c) $y=x^2(x-1)$ (d) $y=x^2(x^2-1)$
(e) $y=(1-x)(2-x)(3-x)$ (f) $y=x^2|x-1|$

6 Sketch the following curves, for values of x, $-360°\leqslant x\leqslant 360°$ showing the greatest and least values of y:

(a) $y=\sin\frac{1}{2}x$ (b) $y=\sin 2x$
(c) $y=\cos^2 x$ (d) $y=1+\cos x$
(e) $y=2\sin(x+120°)$

7 Sketch the following curves, showing the points at which each meets the coordinate axes, and indicate the nature of the curve when x is large:

(a) $y=e^{-x}$ (b) $y=e^{x^2}$
(c) $y=\ln(x^2)$ (d) $y=\ln(1/x)$

8 By first using two rough sketches, sketch the following, showing the points at which each meets the coordinate axes and indicating the nature of the curve when x is large and positive:

(a) $y=x\sin x$ (b) $y=x\ln x$ (c) $y=e^{-x}\sin x$

9 Sketch the following curves, indicating in each the five features listed on p72 where appropriate:

(a) $y=\dfrac{1}{x-2}$ (b) $y=\dfrac{x+1}{x-2}$ (c) $y=\dfrac{x}{x^2+4}$

(d) $y=\dfrac{x}{x^2-4}$ (e) $y^2=x^2(4-x^2)$

PERMUTATIONS AND COMBINATIONS

CONTENTS

ARRANGEMENTS

n objects all different can be arranged in $n!$ different orders.

n objects, p of one kind, q of another, and the rest all different, can be arranged in $\dfrac{n!}{p!q!}$ different ways, e.g., there are $\dfrac{8!}{2!2!2!}$ different ways of arranging the letters in CALCULUS.

SELECTIONS

r objects can be selected from n different objects in $\dfrac{n!}{r!(n-r)!}$ ways, if the order of selection does not matter.

If any number of objects is to be selected from n, all different, that can be done in $2^n - 1$ different ways.

If any number of objects is to be selected from n objects, p of one kind, q of another and the rest all different, that can be done in $(p+1)(q+1)2^{n-p-q} - 1$ different ways.

RELATIONS BETWEEN BINOMIAL COEFFICIENTS

$\dbinom{n}{r} = \dbinom{n}{n\text{-}r}$ formerly $^nC_r = {}^{n-r}C_n$

$\dbinom{n}{r\text{-}1} + \dbinom{n}{r} = \dbinom{n+1}{r}$ formerly $^nC_{r-1} + {}^nC_r = {}^{n+1}C_r$

ARRANGING n OBJECTS IN A LINE

The first object can be chosen in n ways, the second in $(n-1)$ ways, the third in $(n-2)$ ways, and so on, till finally the nth object is the only one left. Thus objects can be arranged in a line in $n(n-1)(n-2)\ldots 3\times2\times1$ ways, i.e. $n!$ ways.

ARRANGING r OBJECTS CHOSEN FROM n

If we wish to select just r objects out of n and arrange them in order, we can choose the first in n ways, the second in $(n-1)$ ways ... and finally the rth in $(n-r+1)$ ways, so that the number of different arrangements in $n(n-1) \ldots (n-r+1)$, i.e. $\dfrac{n!}{(n+r)!}$. This can be written ^nP_r or $_nP_r$.

Example 8.1 In how many ways can a first, second and third prize be awarded to a class of 15 pupils?

The first prize can be awarded in 15 ways, the second in 14 ways and the third in 13 ways. Thus there are $15 \times 14 \times 13$ different ways of awarding the prizes, i.e. 2730.

N.B. $15 \times 14 \times 13 = \dfrac{15!}{12!} = {}^{15}P_{12}$

Example 8.2 In how many ways can 10 persons be seated at a circular table?

In this example, since the table is circular there is no 'head'. Place one person and fix all the others relative to him (or her). The person on his left can be chosen in 9 ways, the person on his left in 8 ways ... so that there are $9 \times 8 \times 7 \ldots \times 1$ ways, i.e. 9!, 362 880.

In general, n persons can be arranged round a circular table in $(n-1)!$ ways; n beads can be threaded on a circular wire in $\frac{1}{2}(n-1)!$ since the two sides of the wire are considered to be indistinguishable.

ARRANGING n OBJECTS, p OF WHICH ARE IDENTICAL

Suppose we have six objects, two of which are identical. Denote the objects by A, B, C, D, X_1 and X_2, suffixes being added to the two otherwise-identical objects denoted by X. Then there are 6! ways of arranging these different objects. But for every arrangement say A X_1 B C X_2 D there is another arrangement A X_2 B C X_1 D that will be indistinguishable when the suffices are removed, so that then the number of different arrangements will be $\frac{1}{2} \times 6!$. More generally, the number of ways of arranging n objects, p of which are identical, is $\dfrac{n!}{p!}$.

Example 8.3 Find the number of ways of arranging 10 different pencils, 4 of which are identical red pencils, and the other six are all different.

Using the result above, the number of different arrangements is

$$\frac{10!}{4!} \text{ i.e. } 10 \times 9 \times 8 \times 7 \times 6 \times 5$$

i.e. 151 200

SELECTING r OBJECTS FROM n

If we choose three objects, say A, B, C from a set of n, selecting them in the order A, B, C, gives us the same final selection as if we select them in the order A, C, B, or the order B, C, A. There are $3 \times 2 \times 1$ different orders in which we can select these three objects; ABC, ACB, BAC, BCA, CAB, CBA, so that the number of different final selections that we can have of three objects from n is

$$\frac{n!}{(n-3)!} \div 6 \text{ i.e. } \frac{n!}{(n-3)!3!}$$

In general, the number of different selections of r objects from n is

$$\frac{n!}{(n-r)!r!}$$

This is now written $\binom{n}{r}$ though the notation $_nC_r$ and ^nC_r may still be met.

SELECTING ANY NUMBER OF OBJECTS FROM n

The first object can either be selected or not selected; likewise the second, and third ... and so on. Thus there are $2 \times 2 \ldots$ different selections, i.e. 2^n. But in one of these, all the objects will have been rejected, i.e. none will have been selected, so the number of different selections is only $2^n - 1$.

Alternatively, the number of ways of selecting 1 object is written $\binom{n}{1}$ etc. We require $\binom{n}{1} + \binom{n}{2} + \binom{n}{3} + \ldots \binom{n}{n}$. Using the binomial theorem to expand $(1+x)^n$

$$(1+x)^n = 1 + \binom{n}{1}x + \binom{n}{2}x^2 \ldots \binom{n}{n}x^n$$

Put $x=1$,

$$2^n = 1 + \binom{n}{1} + \binom{n}{2} + \binom{n}{3} + \ldots + \binom{n}{n}$$

i.e.

$$\binom{n}{1} + \binom{n}{2} + \binom{n}{3} \ldots \binom{n}{n} = 2^n - 1$$

Example 8.4 Find the number of different outcomes of 13 football matches, if each match can end in a home win, a draw or an away win.

This is a variation of the problem of choosing any numbers of objects, as instead of there being two possible outcomes for each object, there are three possible results of each match, so that there are

$$3 \times 3 \times \ldots = 3^{13} \text{ possible outcomes}$$

RELATIONS BETWEEN BINOMIAL COEFFICIENTS

(1)   Since if we select any r objects from n, we automatically reject $(n-r)$ objects,

$$\binom{n}{r}\binom{n}{n-r}$$

This is illustrated by the symmetry of the binomial coefficients in Pascal's triangle.

(2) $\binom{n}{r-1}+\binom{n}{r}=\binom{n+1}{r}$

To choose r objects from $(n+1)$ different objects, denote the objects by $a_1, a_2, \ldots a_n, b$. Then either we choose b or we do not choose b. If we choose b, we have to choose $(r-1)$ objects from the n as, which is done in $\binom{n}{r-1}$ ways; if we do not choose b, we must choose all r objects from the n as, in $\binom{n}{r}$ different ways.

Thus $\binom{n+1}{r}=\binom{n}{r-1}+\binom{n}{r}$

This result is used in the binomial theorem (p43). It can also be proved by writing each binomial coefficient in terms of factorials.

QUESTIONS

1   Find the number of different arrangements of the letters ASTER.
2   Find the number of different arrangements of the letters WALL-FLOWER.
3   Find the number of ways in which 8 different beads can be arranged on a circular wire.
4   Find the number of arrangements on a circular wire of 8 beads, 3 of which are identical red beads and the other 5 are all different.
5   In how many ways can a team of four be chosen from eight players?
6   In how many ways can two teams of four be chosen from eight players?
7   In how many ways can two teams of four be chosen from nine players?
8   In how many ways can a selection of fruit be made up from an apple, a banana, an orange and a pear?
9   In how many ways can a selection of fruit be made up from 2 apples, 3 bananas, 4 oranges and 5 pears?
10 Find the number of selections of three cards chosen from a pack of 52 different playing cards. Find also the number of selections of three cards from a suit of 13 different cards.

11 In how many ways can four boys and four girls be arranged in a circle? In how many of these will boys and girls occur alternately?

12 In how many ways can n boys and n girls be arranged in a circle? In how many of these will the boys and girls occur alternately?

PROBABILITY

CONTENTS

The probability of an event A, is defined as

$$\frac{\text{the number of } \textbf{equiprobable favourable } \text{outcomes}}{\text{total number of } \textbf{equiprobable } \text{outcomes}}$$

If A and B denote two events

$$P(A \text{ or } B \text{ or both}) = P(A) + P(B) - P(\text{both } A \text{ and } B)$$

This can be written

$$P(A \cup B) = P(A) + P(B) - P(A \cap B)$$

DEPENDENT EVENTS

$P(B|A)$ denotes the probability of event B, **given that event A has already happened**.

$$P(B|A) = \frac{P(A \cap B)}{P(A)}$$

Remember that *tree diagrams* are often useful.

BINOMIAL DISTRIBUTION

If the probability of success in one trial is p, the probability of r successes out of n trials is

$$\frac{n!}{r!(n-r)!} \, p^r (1-p)^{n-r}$$

GEOMETRIC DISTRIBUTION

If an experiment finishes as soon as a success is recorded, the probability of a success at the rth attempt is $(1-p)^{r-1} p$.

DEFINITION

The probability of an event A written $P(A)$ is defined as the ratio

$$\frac{\text{the number of equiprobable favourable outcomes}}{\text{the total number of equiprobable outcomes}}$$

e.g. P (card drawn at random from a pack of 52 cards is a king) is 4/52, since there are four kings; P (a fair die shows a number divisible by 3) is 2/6, since there are two numbers, 3 and 6, divisible by three. Note *equiprobable* favourable events. It is incorrect to say that because there are three possible outcomes when throwing a pair of coins, TT, TH and HH, that the probability of both coins showing heads is 1/3; the three possible outcomes are not equiprobable. The correct probability of two heads is 1/4.

MUTUALLY EXCLUSIVE EVENTS

If two events are such that one excludes the other, the events are said to be mutually exclusive, e.g. a coin cannot show heads and tails, a match cannot be both lost and won.

INDEPENDENT EVENTS

If two events are such that neither has any effect on the other, they are called independent events. Unless events are clearly dependent on each other, such as the outcome of drawing a second card from a pack when the first card drawn has not been replaced, assume that the events are independent unless told otherwise. In practice it might be argued that they are not independent, because if one athlete sets a fast pace to break the record then that may help the other athlete, but ignore connections like this unless told otherwise.

DEPENDENT EVENTS, CONDITIONAL PROBABILITY

Two events are dependent if knowledge that one event has occurred affects the probability that the other will occur. If a spade has been drawn from a standard pack of 52 cards, the probability that the next card drawn is a spade is 12/51, because there are only 12 spades left out of 51 cards.

The probability of an event A given that an event B has already happened, is written $P(A|B)$. If two events are independent, so that the probability of the second event is not affected by the first, $P(A|B)=P(A)$, and $P(B|A)=P(B)$

Example 9.1 A card is drawn at random from a standard pack of 52 cards, and then replaced. The probability that the next card is a spade is 13/52, 1/4, and is not affected by whether the first card was a spade or not. The draws are independent. If the first card is not replaced, however, then the probability that the second card is a spade is affected by whether or not the first is a spade, and these events are dependent.

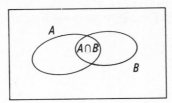

Fig 9.1

When we are calculating $P(B|A)$, we know that A has already occurred, so that A represents the set of all possible outcomes. The event B only occurs if any of the outcomes in $A \cap B$ occur, so that

$$P(B|A) = \frac{n\{A \cap B\}}{n\{A\}} = \frac{P(A \cap B)}{P(A)}$$

ADDITION LAW

If events A and B are mutually exclusive, the probability of both A and B is $P(A) + P(B)$. More generally, if $P(A \cup B)$ is the probability of A and B,

$$P(A \cup B) = P(A) + P(B) - P(A \cap B).$$

This is illustrated by the Venn diagram in Fig 9.2.

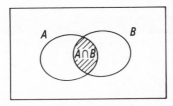

Fig 9.2

Example 9.2 Find the probability that a card drawn at random from a pack of 52 cards is either a two or a spade.

P(two)	=4/52
P(spade)	=13/52
P(two of spades)	=1/52
P(two or a spade)	=4/52×13/52−1/52=16/52

Alternatively, we can see that there are 16 different favourable cards, the thirteen spades, and the three cards, two of hearts, diamonds and clubs, out of a total of 52 different cards.

APPLICATION OF VENN DIAGRAMS

Many problems can be solved using Venn diagrams, as in this example.

Example 9.3 Two events A and B are such that $P(A)=0.2$, $P(A \cap B)=0.15$ and $P(A' \cap B)=0.25$. Find $P(A \cap B')$ and $P(A|B)$.

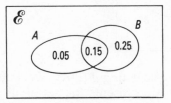

Fig 9.3

Since the probability of any event is proportional to the number of elements in a certain set, we can fill in the Venn diagram as above, starting with $A \cap B=0.15$, the value of 0.05 being $0.2-0.15$. This is $P(A \cap B')$, so

$$P(A \cap B')=0.05$$

To find $P(A|B)$, we have to choose events A from the set B, and we see that

$$P(A|B)=\frac{P(A \cap B)}{P(B)}$$

$$=\frac{0.15}{0.4}=0.375$$

CONDITIONAL PROBABILITY

Sometimes these problems can be solved using a Venn diagram, sometimes by just using the formula

$$P(A|B)=\frac{P(A \cap B)}{P(B)},$$

sometimes it helps if we tabulate the equiprobable outcomes, and use our definition of probability

$$P(A)=\frac{\text{number of equiprobable favourable outcomes}}{\text{total number of equiprobable outcomes}}$$

Example 9.4 Three fair coins are spun. Find the probability that two of them show Heads, given that at least one shows Heads.
   The equiprobable outcomes are

TTT, TTH, THT, HTT, HHT, HTH, THH, HHH

and we have to select from the last seven of these, since we are given that at least one shows Heads. Of these

HHT, HTH, THH

are 'favourable', so that

P(two show Heads|one shows Heads)=3/7

Notice that P(two show heads)=3/8
P(one shows Heads)=7/8
so that P(two show Heads|at least one shows Heads)=$\frac{3/8}{7/8}=\frac{3}{7}$

PRODUCT LAW

If events A and B are independent, then the probability that both events happen, written $P(A \cap B)$ is given by

$$P(A \cap B)=P(A) \times P(B)$$

Example 9.4 The probability that Jack catches the bus one morning is 1/3; the probability that Jill catches the bus is 1/4.

Find the probability that (a) both catch the bus, (b) both miss the bus, (c) just one misses the bus.

(a) P(Jack catches the bus) =1/3
P(Jill catches the bus) =1/4
P(both catch the bus) =1/3×1/4=1/12.

(b) Either Jack catches the bus or he does not, so the probability that he does not catch the bus is 1−1/3, i.e. 2/3; similarly the probability that Jill does not catch the bus is 1−1/4, i.e. 3/4, so the probability that both miss the bus is 2/3×3/4, i.e. 1/2.

(c) If just one misses the bus, then either Jack catches the bus but Jill does not, 1/3×3/4, i.e. 1/4, or Jack does not catch the bus but Jill does, probability 2/3 × 1/4, i.e. 1/6.

∴ P(just one misses the bus)=1/4+1/6=5/12

Wherever possible, check probabilities, and here we see that

P(both catch the bus) =1/12
P(just one catches the bus) =5/12
P(both miss the bus) =1/2

These are all the possible outcomes, and 1/12+5/12+1/2=1.

USE OF TREE DIAGRAMS

A model of the climate in a certain region may be made by classifying each day as wet or fine, and saying that if it is fine one day, the probability that it is fine the next will be 3/4; that if it is wet one day the probability that it will be wet the next is 2/3. If it is fine on Monday of one week, can we find the probability that it will be fine on Wednesday and Thursday?

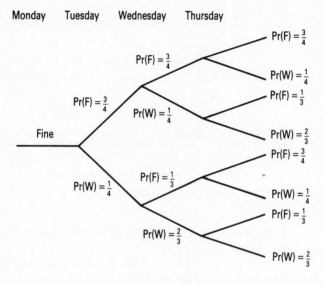

Fig 9.4

Draw a tree diagram (Fig 9.4) and complete the probability of events. Then if it is fine on Wednesday, we can have FF or WF. Thus

P(fine on Wednesday)	$=P(FF)+P(WF)$
	$=3/4\times3/4+1/4\times1/3$
	$=31/48$

Similarly,

P(wet on Wednesday)	$=P(WW)+P(FW)$
	$=1/4\times2/3+3/4\times1/4$
	$=17/48$

and we notice that

31/48+17/48   =1

To find the probability that it is fine on Thurday, we have

P(fine on Thursday)	$=3/4\times$P(fine on Wednesday)
	$+1/3\times$P(wet on Wednesday)
	$=3/4\times31/48+1/3\times17/48$
	$=347/576$

and

P(wet on Thursday)	$=1/4\times$P(fine on Wednesday)
	$+2/3\times$P(wet on Wednesday)
	$=1/4\times31/48+2/3\times17/48$
	$=229/576$

Notice again that

$$347/576 + 229/576 = 1$$

BINOMIAL DISTRIBUTION

To find the probability of one failure (F) out of four trials, favourable outcomes can be written

FSSS; SFSS; SSFS and SSSF

Thus if the probability of success is p, the probability of failure is $(1-p)$, and the probability of the outcome FSSS is $(1-p)p^3$; since there are four equiprobable favourable outcomes, each with a probability $(1-p)p^3$, the probability of one failure out of four trials is $4(1-p)p^3$. Generally, if we have an experiment in which we want r successes out of n trials, there will be $\binom{n}{r}$ ways in which these r successes can occur, each with a probability of $p^r(1-p)^{n-r}$, so that the probability of r successes out of n trials is $\binom{n}{r}p^r(1-p)^{n-r}$. This is an example of a binomial distribution.

Example 9.5 When practising, the probability that a tennis player serves an 'Ace' is 0.2. Find the probability that he serves an 'Ace' twice in five attempts.

The probability p that he serves an 'Ace' is 0.2, so the probability that he does not serve an 'Ace' is 0.8. If A denotes that he serves an 'Ace' and X that he does not serve an 'Ace', then AAXXX would be an acceptable order, so would AXAXX and AXXAX. There are $\frac{5!}{2!3!}$ i.e. $\binom{5}{2}$, 10 such orders. The probability of each is $(0.2)^2(0.8)^3$, so that the probability that he serves two 'Aces' out of five attempts is $10(0.2)^2(0.8)^3$, i.e. 0.2048.

GEOMETRIC DISTRIBUTION

Some experiments end as soon as one success (or failure) has been attained. A cricketer's innings finishes as soon as he is 'out', many games begin by throwing a 'six', and most students stop when they have found one solution to a problem, instead of trying to find a better solution. If we suppose that an experiment finishes after a success, then possible outcomes are

S, FS, FFS, FFFS, ...

every experiment finishing as soon as a success is recorded. Denoting the probability of success by p, the probability of the experiment finishing after five failures and one success is $(1-p)^5 p$, after r failures and one success by $(1-p)^r p$. This is an example of geometric distribution.

Example 9.6 The probability that a certain person passes the driving test is 0.7. Find the probability of that person passing (a) at the second attempt (b) at the tenth attempt.

Since the probability of passing is 0.7, the probability of failing is 0.3, so the probability of passing (S) at the second time is $P(F) \times P(S) = 0.3 \times 0.7 = 0.21$.

The probability of passing at the tenth attempt is the probability of nine failures then success, i.e. $(0.3)^9 (0.7)$, about 1.4×10^{-5}.

QUESTIONS

1 Events A and B are such that $P(A)=0.7$, $P(B)=0.5$, and $P(A \cap B)=0.3$. Find   (a) $P(A \cup B)$ (b) $P(A|B)$ (c) $P(B|A)$.

2 Events A and B are such that $P(A)=0.7$, $P(B)=0.4$, and $P(A \cup B)=0.9$. Find   (a) $P(A \cap B)$ (b) $P(A|B)$ (c) $P(B|A)$.

3 Events A and B are such that $P(A)=0.7$, $P(B)=0.4$, and $P(A \cap B)=0.28$. Find   (a) $P(A|B)$ (b) $P(A|B')$ (c) $P(B|A)$.

4 Events A and B are such that $P(A)=a$, $P(B)=b$. Prove that $P(A \cap B)=ab$ if and only if $P(A|B)=P(A|B')$.

5 Events A and B are such that $P(A' \cap B)=0.15$, $P(A \cap B')=0.2$, and $P(A|B)=0.4$. Find   (a) $P(A \cap B)$ (b) $P(A \cup B)$ (c) $P(B|A)$

6 The probability that a card drawn at random from a pack is 1/4. Find the probability that, when four draws are made,
(a)   three are spades,
(b)   three are spades, given that at least one is a spade,
(c)   three are spades, given that at least two are spades,
(d)   three are spades, given that at least three are spades.

7 With the data of Q.6, find the probability that, when four draws are made,
(a)   two are spades, given that at least one is a spade,
(b)   one is a spade, given that at least one is a spade.

8 The probability that the bus is late any one day is 0.2. Find the probability that, in a week of five days,
(a)   the bus is late on two days,
(b)   it is late on two days, given that it is late on at least one day,
(c)   it is late on two days, given that it is late on at least two days.

9 In a certain examination, 80% of the candidates pass English and 75% pass French, whereas 15% fail both examinations. Find the probability that a candidate selected at random
(a)   passes both examinations,
(b)   passes both examinations, given that he passes English,
(c)   passes both examinations, given that he passes French.

10 In a certain village, 50% of the inhabitants have cars, 60% have

bicycles and 30% have neither. Find the probability that a person selected at random

(a)   has both a car and a bicycle, given that he has a car,

(b)   has both, given that he has a bicycle.

11   Red and yellow dice are rolled and the numbers on their faces are recorded. Find the probability that these numbers

(a)   total 2   (b)   total 7   (c)   total more than 7.

Given that the red die shows a 4, find the probability of each of the events above.

12   Two numbers, not necessarily different, are chosen at random from the numbers 1, 2, 3, ... 10. Find the probability that these numbers

(a)   are both even   (b)   are the same   (c)   are different

Given that the first number is 3, find now the probability of each of the above events.

13   The probability that Arthur solves a particular problem is 1/2; that Bert solves that problem is 2/5; that Chris solves the problem is 3/4. Given that these probabilities are independent, find the probability that

(a)   all three solve the problem,

(b)   only one solves the problem,

(c)   the problem is not solved by any of them.

14   From a standard pack of cards all the spades and all the aces are removed. Find the probability that

(a)   a card drawn at random is the queen of diamonds,

(b)   when two cards are drawn together at random, both are queens,

(c)   when two cards are drawn together at random, neither is a queen.

15   If a boy oversleeps one morning, the probability that he oversleeps the next day is 0.1; if he wakes on time one morning, the probability that he wakes on time the next is 0.3. If he wakes on time on Monday, find the probability that he

(a)   wakes on time on Wednesday,

(b)   oversleeps on Wednesday,

(c)   oversleeps on Thursday,

(d)   wakes on time on Friday.

16   The probability of success in a certain trial is 0.4. Find the probability that in 5 trials, there are

(a)   5 successes   (b)   4 successes

(c)   3 successes   (d)   2 successes

(e)   1 success   (f)   no successes.

17   The probability that a certain biassed coin shows 'Heads' is 0.8. Find the probability that in 5 throws,

(a)   there are 3 'Heads',   (b)   there are 4 'Heads',

(c)   there are more 'Heads' than 'Tails'.

18   In a certain district, the weather is classified as either 'dry' or 'wet', and the weather any one day is independent of the weather on any other day. The probability that any one day is 'wet' is 0.3. Find the probability that in a week of 7 days,

(a)   there are 5 'wet' days,

(b)   there are at least 5 'wet' days,

(c)   there are less than 5 'wet' days.

19   A certain 'one armed bandit' costs 5p a time, but pays out £1 if more than 8 balls out of 10 roll into a specified slot. If the probability of any one ball rolling into that slot is 0.6, find the probability that

(a)   all 10 balls roll there,   (b)   no balls roll there,

(c)   only nine of the ten balls roll there,

(d)   only one of the ten balls rolls there,

Deduce the probability of a player winning with one attempt.

20   Some of the tins of one brand of baked beans contain sausages, but these tins are indistinguishable from those that do not contain sausages. The probability that any one tin opened at random contains a sausage is 0.3. Find the probability that out of 10 tins

(a)   none contain sausages, (b)   all ten contain sausages,

(c)   exactly two contain sausages,

(d)   exactly two do not contain sausages.

21   The probability of success in a certain trial is 0.6. Find the probability that the first success is

(a)   at the first attempt,

(b)   at the second attempt,

(c)   at the third attempt,

(d)   before the fourth attempt.

22   The probability that a golfer holes a putt is 0.3. Find the probability that, when practising, he holes his first putt

(a)   at the first attempt,

(b)   at the second attempt,

(c)   at the third attempt.

23   How many attempts will the golfer in Q.22 need to have a probability of 0.9 of having holed his putt?

24   A schoolmaster has bought a large consignment of tins of cat food and dog food at a sale. Although the tins have all lost their labels, he is assured that three-quarters of them contain cat food. To feed his cat, he has to open tins successively until he comes to one containing cat food. Find the probability that

(a)   the second tin he opens contains cat food,

(b)   the fourth tin he opens contains cat food,

(c)   he opens five or more tins before finding one with cat food.

25   Two friends A and B play a game, throwing a fair die alternately, the first to throw a '5' winning. A throws first. Find the probability that he wins.

26   The school chess champion plays two opponents A and B in turn until one of them beats him. The probability that the champion beats A is 0.9; that the champion beats B is 0.8. If the first game is against A, find the probability that it is A who beats the champion.

STRUCTURE

CONTENTS

NOTES

A set S is **closed** under an operation $*$ if $a*b \in S$ for all $a, b \in S$.

An operation is **associative** if $a*(b*c)=(a*b)*c$ for all $a, b, c \in S$.

An operation is **commutative** if $a*b=b*a$ for all $a, b \in S$.

The **identity** element e is such that $a*e=e*a=a$ for all elements $a \in S$.

The **inverse** of an element a, written a^{-1}, is such that $a*a^{-1}=a^{-1}*a=e$.

A set S is a **group** under an operation $*$ if
(a) S is closed under $*$,
(b)   the operation $*$ is associative over S,
(c)   there is an identity element e,
(d)   for every element $a \in S$, there is an inverse $a^{-1} \in S$.

Commutative groups are called **Abelian** groups.

BINARY OPERATIONS

A binary operation is an operation defined on two elements of a set, e.g. 'add 2 to 3', 'subtract 2 from 6'. A binary operation is often denoted by $*$, and may be defined in words, as above, or algebraically.

Example 10.1 A binary operation $*$ is defined over $\mathbf{R}$, the set of all real numbers, by $x*y=2x-y$. Find $x*y$ if $x=3$ and $y=4$.

$$\text{Since } x*y=2x-y$$
$$3*4=6-4$$
$$=2$$

CLOSURE

When a binary operation $*$ is defined over a set S, the set S is closed under $*$ if $a*b$ belongs to S, for all $a, b \in S$. Thus the set $\mathbf{R}$ is closed under the binary operation 'add', for the sum of any two real numbers is a real number. The set $\mathbf{Z}$ of integers is closed under 'add', for the sum of any two integers is an integer. However, $\mathbf{Z}$ is not closed if, say $x*y=x/y$, for if $x=5$ and $y=3$, both elements of $\mathbf{Z}$, $x*y=5/3$, which is not an element of $\mathbf{Z}$.

Example 10.2 (*a*)The set of all even numbers is closed under addition, subtraction and multiplication, but not under division e.g. 6/2=3.

(*b*) The set of all positive integers is closed under addition and multiplication, but not if $x*y=x-y$, e.g. $3-7$ is negative, nor if $x*y=x/y$, for 3/7 is not an integer.

(*c*) The set of all powers of 2 is closed under multiplication and if $x*y=x/y$, but not under addition, nor if $x*y=x-y$.

Notice that one counter-example is sufficient to prove that a set is not closed under a particular operation. The proofs that in certain circumstances the sets are closed, are left as an exercise.

COMMUTATIVE OPERATIONS

If the order in which we combine the elements does not affect the result, i.e. $x*y=y*x$ for all x, y, the operation is called **commutative**. Thus addition is commutative, for $x+y=y+x$, but subtraction is not commutative, for $x-y\neq y-x$ for x, y. Notice that above, we could describe an operation as 'addition', but not, strictly, as subtraction, for the word does not say whether $x*y=x-y$ or $y-x$.

ASSOCIATIVE

If we wish to combine three elements under a binary operation, we have to combine them in pairs. Thus to find $1+2+3$ we find either $(1+2)+3$ or $1+(2+3)$. If the order is immaterial, as in this example, the operation is associative, but notice that, e.g. 'subtraction' is not associative, for $1-(2-3)\neq(1-2)-3$. If an operation is associative,

$$a*(b*c)=(a*b)*c$$

for all a, b and c.

Example 10.3 A binary operation $*$ is defined over $\mathbf{R}$ by $x*y=\frac{1}{2}(x+y)$. Investigate whether $\mathbf{R}$ is closed under $*$, and whether $*$ is associative or commutative.

If x and y are any real numbers, $\frac{1}{2}(x+y)$ is also a real number, so $\mathbf{R}$ is closed under $*$. Notice though that if the same operation is defined over $\mathbf{Z}$, it is not closed, for $\frac{1}{2}(2+1)$ is not an integer.

Since $\frac{1}{2}(x+y)=\frac{1}{2}(y+x)$, the operation is commutative, but $x*(y*z)=x*\frac{1}{2}(y+z)=\frac{1}{2}x+\frac{1}{4}y+\frac{1}{4}z$ whereas $(x*y)*z=\frac{1}{4}(x+y)+\frac{1}{2}z$, so the operation is not associative.

IDENTITY ELEMENT

When zero is added to any number, the value of the number is not altered. When any number is multiplied by one, the number is not altered. 0 is called the identity element under addition; 1 is the identity element under multiplication. The identity element (usually denoted by e) has the property that $a*e=e*a=a$ for all a.

INVERSE ELEMENT

$4\times\frac{1}{4}=1$ and $\frac{1}{4}\times4=1$, 1 being the multiplicative identity under multiplication. $\frac{1}{4}$ is called the multiplicative inverse of 4 (and 4 is the multiplicative inverse of $\frac{1}{4}$). The inverse of an element a(usually denoted by a^{-1}) is the element such that $a*a^{-1}=a^{-1}*a=e$. Thus the additive inverse of 4 is -4, since $4+(-4)=(-4)+4=0$, 0 being the identity element under addition.

Example 10.4 Multiplication mod 5 is defined over the set of integers S $\{1,2,3,4\}$. Show that S is closed under this operation. Find the identity element, and the inverse of each element.

The table below shows the result of $x*y$, for all $x, y \in S$

	1	2	3	4
1	1	2	3	4
2	2	4	1	3
3	3	1	4	2
4	4	3	2	1

From this we see that $x*y$ belongs to S, for all $x, y \in S$, so that the set is closed under $*$. We see that 1 is the element that leaves all others unaltered, i.e. $x*1=1*x=x$, for all x, so that 1 is the identity element under $*$.

We see that $2*3=3*2=1$, so that the inverse of 2 is 3, and the inverse of 3 is 2. However, $1*1=1$ and $4*4=1$, so that the inverse of 4 is 4, and the inverse of 1 is 1, i.e. 1 and 4 each is its own inverse.

GROUP: DEFINITION

A set of elements S is a group under an operation $*$ if four conditions are satisfied:
(a) S is closed under $*$,
(b)   the operation $*$ is associative over S, i.e. $a*(b*c)=(a*b)*c$ for all $a, b, c \in S$,
(c)   there is an identity element $e \in S$, i.e. an element e such that $a*e=e*a=a$ for all $a \in S$,
(d)   every element $a \in S$ has an inverse a^{-1} in S.

If in addition the operation ∗ is commutative, then the group is called a commutative (or Abelian) group.

Notice that in the example above, all these conditions are satisfied, so that the numbers 1, 2, 3, 4 form a group under multiplication mod 5.

ORDER OF AN ELEMENT; ORDER OF A GROUP

An element a of S is said to have order n if n is the smallest positive integer such that $a^n=e$, where $a^2=a*a$, $a^3=a^2*a=a*a^2$ and a^n is defined by $a^n=a*a^{n-1}$, etc. Thus in a group $\{1, 2, 3, 4\}$ under multiplication mod 5,

> 1 has order 1,
> 2 has order 4, since $2^4=1$,
> 3 has order 4, since $3^4=1$
and   4 has order 2, since $4^2=1$

The number of elements in a group is called the order of the group.

CYCLIC GROUPS, GENERATORS

In a cyclic group, all the elements are 'powers' of one element, e.g. the rotations R through 60° in a plane, where the operation ∗ is 'first one rotation, then the other', $R^2=R*R$, a rotation of 120°, $R^3=R*R^2$, a rotation of 180°, and so on. The element R, corresponding to a rotation of 60°, is called a **generator** of the group, for all other elements can be expressed as powers of R, and the group consists of the six elements R, R^2, R^3, R^4, R^5 and R^6, the last being the identity element. Notice that R^5 would also serve to generate this group, but none of the other elements is a generator. Only cyclic groups have generators.

Example 10.5 Show that the numbers 1, −1, i, −i form a group under multiplication.

Make up a table as below.

	1	−1	i	−i
1	1	−1	i	−i
−1	−1	1	−i	i
i	i	−i	−1	1
−i	−i	i	1	−1

Taking the group axioms in turn we see
(a)   the set is closed under multiplication,
(b)   the operation is associative, though this is tedious to check and is often given in examinations,

(*c*)   there is an identity element, 1,

(*d*)   every element has an inverse, *viz.* $1^{-1}=1$, $(-1)^{-1}=-1$, $(i)^{-1}=-i$ and $(-i)^{-1}=i$

All four criteria are satisfied, so the numbers form a group under multiplication. Notice that *i* and $-i$ are generators of this group, but not 1 or -1.

ISOMORPHIC GROUPS

Two groups are said to be isomorphic if they have essentially the same structure, which can best be seen, for finite groups, by looking at the group table. If, by renaming the elements and possibly altering the order of the rows and columns in one table, the two tables become identical, then the groups are isomorphic.

Consider the numbers {0, 1, 2, 3} under addition mod 4. Then the table is that given in Table 1.

	0	1	2	3
0	0	1	2	3
1	1	2	3	0
2	2	3	0	1
3	3	0	1	2

Table 1

The group table for the numbers 1, 2, 3, 4 under multiplication mod 5 is given in Table 2(*a*). We can see that by altering the order in which the elements are written the table becomes of the same structure as that in Table 2(*b*). By contrast, Table 3 cannot be rearranged to take

	1	2	3	4
1	1	2	3	4
2	2	4	1	3
3	3	1	4	2
4	4	3	2	1

Table 2(*a*)

	1	2	4	3
1	1	2	4	3
2	2	4	3	1
4	4	3	1	2
3	3	1	2	4

Table 2(*b*)

the same form as the first group table, and so is not isomorphic to the first group. Notice also there are no generators of this group.

	e	*a*	*b*	*c*
e	*e*	*a*	*b*	*c*
a	*a*	*e*	*c*	*b*
b	*b*	*c*	*e*	*a*
c	*c*	*b*	*a*	*e*

Table 3

The group to which the integers 0, 1, 2, 3 under addition mod 4 belong is called the cyclic group of order 4; the group illustrated in

Table 3 is called the Klein four-group (after F. Klein, 1849–1925). It can be shown that all groups of order 4 are isomorphic to one or other of these groups. The difference in the structure of these two groups can be seen by noticing that in the Klein four-group, each element is its own inverse, so that all elements are of order 2, whereas in the cyclic group there are two elements of order 4; we have already commented that there are no generators of the Klein four-group.

GROUP OF SYMMETRIES, GROUP OF PERMUTATIONS

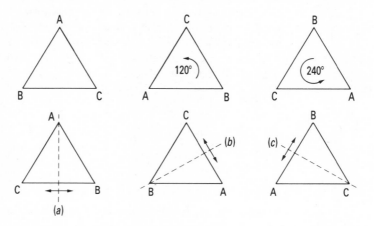

Fig 10.1

The symmetry of an equilateral triangle is such that the triangle is unaltered when rotated in its own plane through 120°. This rotation can be described by the permutation ABC→BCA, in which each letter 'moves on' one place. Similarly, the triangle can be rotated through 240°, described by ABC→ACB, each letter 'moving on' two places. The triangle can also be reflected in axis (a), this reflection being described by ABC→ACB, in which the first letter is unaltered, and the other two interchanged. Similar reflections are described by ABC→CBA (reflection in axis (b)) and ABC→BAC (reflection in (c)), each corresponding to one of the symmetries of the triangle. Let us adopt the following notation. The permutation (which corresponds, as each does, to one of the symmetries of the triangle) ABC→BCA we denote by T_1,

the permutation ABC→CAB we denote by T_2
$$ABC→ACB \qquad T_3$$
$$ABC→CBA \qquad T_4$$
$$ABC→BAC \qquad T_5$$
and the 'identity permutation' ABC→ABC T_0

Then the table below shows that these permutations (and of course the corresponding symmetries) form a group under the operation 'first one, then the other':

	T_0	T_1	T_2	T_3	T_4	T_5
	ABC	BCA	CAB	ACB	CBA	BAC
T_0	T_0	T_1	T_2	T_3	T_4	T_5
T_1	T_1	T_2	T_0	T_4	T_5	T_3
T_2	T_2	T_0	T_1	T_5	T_3	T_4
T_3	T_3	T_5†	T_4	T_0	T_2††	T_1
T_4	T_4	T_3	T_5	T_1	T_0	T_2
T_5	T_5	T_4	T_3	T_2	T_1	T_0

where, for example, the element marked †, T_3*T_1 means 'first T_1 then T_3' and describes ABC→BCA→BAC, equivalent to the single permutation T_5 (corresponding to a reflection in (c)) and the element marked †† T_3*T_4 means 'first T_4 then T_3' and describes ABC→CBA→CAB, equivalent to T_2. Notice that the table is not symmetrical about the leading diagonal, showing that the group is not a commutative (i.e. Abelian) group.

We see in the table above that the set $\{T_0, T_1, T_2\}$ is closed, and is an example of a *subgroup*. Other subgroups are $\{T_0, T_3\}$, $\{T_0, T_4\}$ and $\{T_0, T_5\}$. The order of the original group of 6; of the subgroups 3, 2 and 2 and 2 respectively. It is not a coincidence that these are factors of 6, but the proof of this (Lagrange's theorem) is outside present syllabuses.

QUESTIONS

1 The operation $*$ is defined over **R** by $x*y=1/x+1/y$. Is **R** closed under this operation? Is the operation associative or commutative?
   The same operation $*$ is now defined over **Z**. Is **Z** closed under$*$?

2 The operation $*$ is defined over $\mathbf{R}^+$ by $x*y=\sqrt{xy}$, the positive square root being taken. Is $\mathbf{R}^+$ closed under$*$? If the operation is defined over **Z**, is **Z** closed under $*$?
   Investigate whether $*$ is associative or commutative.

3 If $A=\begin{pmatrix} 0 & 1 \\ -1 & 0 \end{pmatrix}$ show that the set $\{A, A^2, A^3, A^4\}$ forms a group under matrix multiplication, and that this group is isomorphic to that formed by $\{1, -1, i, -i\}$ under multiplication.

4 Show that the numbers $\{1, 3, 5, 7\}$ form a group under multiplication mod 8. Give the inverse of each element, the order of each element, and say to which of the groups of order four it is isomorphic.

5 Say why each of the following does not form a group:
   (a)   the numbers $\{1, 2, 3\}$ under multiplication mod 4,
   (b)   the real numbers under the operation $*$ where
          $a*b=a+b-ab-1$,

(c)   the real numbers under the operation ~, where ~ means the positive difference,

e.g.   7~2=5 and 2~7=5.

6   If a and b are any elements in any group, show that the inverse of $(a*b)$, written $(a*b)^{-1}$, is $b^{-1}*a^{-1}$.

7   Functions $f_1, \ldots$ are defined by

$$f_1:x\rightarrow x, \; f_2:x\rightarrow\frac{1}{x}, \; f_3:x\rightarrow -x \text{ and } f_4:x\rightarrow -\frac{1}{x}.$$

Show that these form a group under the usual rule for composition of functions, and find to which group of order four it is isomorphic.

8   The symmetries of a rectangle ABCD (Fig 10.2)

are   rotation through 180°, described by ABCD→CDAB

reflection in Ox ABCD→DCBA

reflection in Oy ABCD→BADC

and   the identity element   ABCD→ABCD

Draw up a table to show that these form a group of order four. To which of the two groups, the cyclic group or the Klein four-group, is this isomorphic?

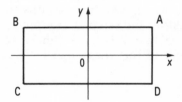

9   Describe the symmetries of the rhombus in the same manner as we have described the symmetries of the rectangle. Show that these form a group of order four, and find to which group this is isomorphic.

RECTANGULAR CARTESIAN COORDINATES, STRAIGHT LINE AND CIRCLE

CONTENTS

NOTES

If P_1 has coordinates (x_1, y_1) etc., the **distance between the two points P_1 and P_2** is

$$\sqrt{[(x_1-x_2)^2+(y_1-y_2)^2]}$$

the **midpoint** of $P_1 P_2$ is

$$\tfrac{1}{2}(x_1+x_2), \tfrac{1}{2}(y_1+y_2)$$

the coordinates of the **point dividing $P_1 P_2$ in the ratio** $\lambda:\mu$ are

$$\frac{\mu x_1+\lambda x_2}{\lambda+\mu}, \frac{\mu y_1+\lambda y_2}{\lambda+\mu}$$

the **centroid** of the triangle $P_1 P_2 P_3$ is

$$\tfrac{1}{3}(x_1+x_2+x_3), \tfrac{1}{3}(y_1+y_2+y_3)$$

the equation of the **straight line through P_1, P_2** is

$$\frac{y-y_1}{y_2-y_1} = \frac{x-x_1}{x_2-x_1}$$

the equation of the **straight line through P_1, with gradient m** is

$$y-y_1=m(x-x_1)$$

Two line gradients m_1, m_2 are **parallel** if $m_1=m_2$, are **perpendicular** if $m_1 m_2=-1$

The **equation of the circle** centre $(h, k,)$ radius r is

$$(x-h)^2+(y-k)^2=r^2$$

The circle equation $x^2+y^2+2gx+2fy+c=0$ has

centre $(-g, -f)$, **radius** $\sqrt{[g^2+f^2-c]}$

The equation of the **tangent** at (x', y') to the circle $x^2+y^2=r^2$ is
$$xx'+yy'=r^2$$

The equation of the **tangent** at $(x'y')$ to $x^2+y^2+2gx+2fy+c=0$ is
$$xx'+yy'+g(x+x')+f(y+y')+c=0$$

THE DISTANCE BETWEEN TWO POINTS

By Pythagoras' theorem, from Fig 11.1, we see that the distance between the points $P_1(x_1, y_1)$ and $P_2(x_2, y_2)$ is given by

$$P_1P_2 = \sqrt{[(x_1-x_2)^2+(y_1-y_2)^2]}$$

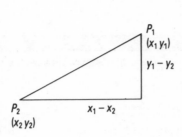

Fig 11.1

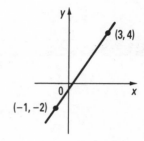

Fig 11.2

Example 11.1 The distance between the points $(-1, -2)$ and $(3, 4)$ is given by

$$(P_1P_2)^2 = [3-(-1)]^2+[4-(-2)]^2$$

i.e.

$$P_1P_2 = \sqrt{52}$$

N.B. It is usually wise to draw a diagram; be very careful of the signs of the numbers.

MIDPOINT OF THE LINE JOINING TWO POINTS

If the coordinates of the midpoint M(Fig 11.3) are (X, Y),

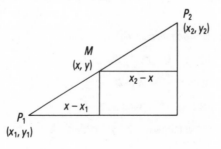

Fig 11.3

considering the X coordinate,

$$X-x_1=x_2-X$$

i.e.
$$X=\tfrac{1}{2}(x_1+x_2)$$

Similarly,
$$Y=\tfrac{1}{2}(y_1+y_2)$$

POINT DIVIDING A LINE IN THE RATIO λ:μ

Considering Fig 11.3 again, if instead of being the midpoint, M divides P_1P_2 in the ratio $\lambda:\mu$,

$$\frac{X-x_1}{x_2-X}=\frac{\lambda}{\mu}$$

whence $\qquad X=\frac{\lambda x_2+\mu x_1}{\lambda+\mu}$

N.B. Note the order of x_2 and x_1. If M divides P_1P_2 in the ratio $\lambda:\mu$, then λ multiplies the coordinates of P_2, μ the coordinates of P_1.

CENTROID OF A TRIANGLE

If A' is the midpoint of BC, the coordinates of A' are

$$\tfrac{1}{2}(x_1+x_2),\ \tfrac{1}{2}(y_1+y_2)$$

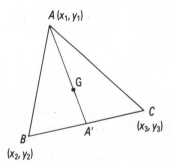

Fig 11.4

If G divides AA' in the ratio 2:1, the x-coordinate of G is

$$\frac{2[\tfrac{1}{2}(x_1+x_2)]+x_3}{2+1},\ \text{i.e. } \tfrac{1}{3}(x_1+x_2+x_3)$$

Similarly, the y-coordinate of G is $\tfrac{1}{3}(y_1+y_2+y_3)$.

By symmetry, the point G defined in this way divides BB' in the ratio 2:1, and divides CC' in the ratio 2:1.

EQUATION OF STRAIGHT LINE

The gradient of any straight line through the points (x_1, y_1) (x_2, y_2) is

$$\frac{y_2-y_1}{x_2-x_1}$$

If P is any point on the straight line P_1P_2, then the gradients of PP_1 and P_2P_1 are equal, i.e.

$$\frac{y-y_1}{x-x_1}=\frac{y_2-y_1}{x_2-x_1}$$

which is conveniently written

$$y-y_1=\frac{y_2-y_1}{x_2-x_1}(x-x_1)$$

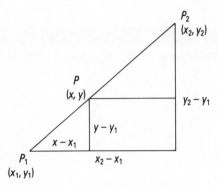

Fig. 11.5

STRAIGHT LINE WITH GIVEN GRADIENT m

If we require the equation of the straight line through the point (x_1, y_1) with gradient m, any point (x, y) on the line has the property that the line joining it to (x_1, y_1) has gradient m,

i.e. $\quad \dfrac{y-y_1}{x-x_1}=m$

i.e. $\quad y-y_1=m(x-x_1)$

STRAIGHT LINE IN INTERCEPT FORM

If the straight line makes intercepts a and b on the x and y axes respectively, it passes through the points $(a, 0)$, $(0, b)$. The equation reduces to the form

$$\frac{x}{a}+\frac{y}{b}=1$$

PARALLEL LINES AND PERPENDICULAR LINES

If two straight lines are parallel, their gradients m_1 and m_2 are equal, i.e. $m_1=m_2$.

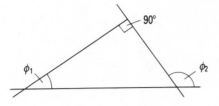

Fig 11.6

If two straight lines are perpendicular, i.e. if $\varphi_2=90°+\varphi_1$ (Fig 11.6), $\tan(\varphi_2-\varphi_1)$ is infinite. But

$$\tan(\varphi_2-\varphi_1)=\frac{\tan\varphi_2-\tan\varphi_1}{1+\tan\varphi_2\tan\varphi_1}$$

i.e. the lines are perpendicular if and only if $\tan\varphi_2\tan\varphi_1+1=0$,

i.e $\qquad m_1m_2=-1$

ANGLE BETWEEN TWO STRAIGHT LINES

Generalizing the result above, if the angle between the two straight lines is α, not necessarily equal to $90°$, then

$$\tan\alpha=\tan(\varphi_2-\varphi_1),$$

$$=\frac{m_2-m_1}{1+m_1m_2}$$

EQUATION OF A CIRCLE

From the definition of a circle, if the point $P(x,y)$ lies on a circle centre (h,k) radius r.

$$(x-h)^2+(y-k)^2=r^2 \qquad (1)$$

In particular, the equation of the circle centre $(0,0)$ radius r is

$$x^2+y^2=r^2$$

Equation (1) can be rearranged as

$$x^2+y^2-2hx-2ky+h^2+k^2-r^2=0$$

so that if the equation is given in the form

$$x^2+y^2+2gx+2fy+c=0 \qquad (2)$$

we can deduce the centre is $(-g, -f)$ and the radius r is given by

$$h^2 + k^2 - r^2 = c, \text{ i.e. } r = \sqrt{(g^2 + f^2 - c)}$$

Example 11.2 The equation of the circle centre $(3, -1)$ radius 4 is

$$(x-3)^2 + (y+1)^2 = 4^2$$

i.e. $\qquad x^2 + y^2 - 6x + 2y - 6 = 0$

Example 11.3 Find the centre and radius of the circle

$$4x^2 + 4y^2 - 8x + 2y = 1$$

First divide by the coefficient of x^2:

$$x^2 + y^2 - 2x + \tfrac{1}{2}y = \tfrac{1}{4}$$

Comparing with (2),

the centre is $(1, -\tfrac{1}{4})$, the radius is $\sqrt{\{1^2 + (-\tfrac{1}{4})^2 + \tfrac{1}{4}\}}$, i.e. $\tfrac{1}{4}\sqrt{21}$

EQUATION OF THE TANGENT AT (x', y')

The gradient at any point (x', y') can be found by differentiating, i.e.

$$2x + 2y\frac{dy}{dx} + 2g + 2f\frac{dy}{dx} = 0$$

whence $\qquad \dfrac{dy}{dx} = -\dfrac{(x'+g)}{(y'+f)}$ at (x', y')

and the equation of the tangent is

$$y - y' = -\left(\frac{x'+g}{y'+f}\right)(x-x')$$

i.e. $\qquad xx' + yy' + g(x+x') + f(y+y') + c = 0 \qquad\qquad (3)$

Example 11.4 Find the equation of the tangent to the circle $x^2 + y^2 + 4x + 6y - 7 = 0$ at the point $(2, -1)$.

Substituting in (3), we have

$$2x - y + 2(x+2) + 3(y-1) - 7 = 0$$

i.e. $\qquad\qquad\qquad\qquad 2x + y - 3 = 0$

Write down the following:

1 The distance between the points (7, 4) and (4, 0).
2 The distance between the points (7, 4) and (−5, −1).
3 The coordinates of the midpoint of the line joining (3, 1) and 5, −3).
4 The coordinates of the midpoint of the line joining (−5, 1) and (4, 1).
5 The coordinates of the point dividing (4, 1), (7, 7) in the ratio 2:1.
6 The coordinates of the point dividing (4,1) and (7, 7) in the ratio 2: −1; represent these points on a diagram.
7 The centroid of the triangle vertices (2, 1), (4,3) and (3, −1).
8 The centroid of the triangle vertices (4, 5), (−1, −4) and (−3, −1).

Write down the equation of the following straight lines:

9 Through (1, 3) and (−2, 4).
10 Through (1, −1) with gradient 2
11 Through (2, 0) and (0, 3)
12 Through (2, −1) inclined at 45° to the x-axis.
13 Through (1, −2) parallel to $3y=2x-4$.
14 Through (1, −1) perpendicular to $3y=2x-4$.

Find the tangent of the angle between:

15 $y=3x-4$ and $y=2x+1$.
16 $3y=x+4$ and $2y+x=1$.
17 Write down the equation of the circle centre (−3, 2) radius 4.
18 Find the coordinates of the centre and the radius of the circle $x^2+y^2-6x+2y-15=0$
19 Find the equation of the circle radius 3 concentric with $x^2+y^2-5x-7y=1$
20 Find the equation of the tangent to $x^2+y^2=25$ at the point (−3, 4).
21 Find the equation of the tangent to $x^2+y^2-6x+2y+6=0$ at the point (3, 1).
22 Find the point on the circle $x^2+y^2-16x+12y+75=0$ which is
    (a)   nearest to
    (b)   furthest from, the origin.
23 Find the equation of the circle through the points (0, 1), (1, 2) (3, 1).
24 Find the equations of the circles touching both coordinate axes and passing through the point (2, 1)
25 Find the two values of m for which the line $my=11-3x$ is a tangent to the circle
$$x^2+y^2-8x-12y+25=0$$
26 Find the equations of the two tangents from the origin to the circle
$$(x-3)^2+(y-2)^2=1$$

PARABOLA AND RECTANGULAR HYPERBOLA

CONTENTS

122   Contents

The locus of a point that moves in a plane so that its distance from a fixed point S, the focus, is equal to its distance from a fixed straight line, the directrix, is a **parabola**.

If the axes are chosen so that the equation is $y^2=4ax$, the coordinates of the **focus** are $(a, 0)$; the equation of the **directrix** is $x=-a$.

The **tangent** to this curve with **given gradient** m is

$$y=mx+a/m$$

the tangent at a **given point** (x', y') is

$$yy'=2a(x+x')$$

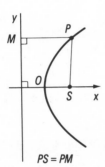

Fig 12.1

$PS = PM$

If the axes are chosen so that the equation of a rectangular hyperbola is $xy=c^2$, the coordinates of the **foci** are $(c\sqrt{2}, c\sqrt{2})$ and $(-c\sqrt{2}, -c\sqrt{2})$; the **directrices** are $x+y=c\sqrt{2}$ and $x+y=-c\sqrt{2}$

The ratio $PS:PM$ is called the **eccentricity** e of a conic; the eccentricity of a rectangular hyperbola is $\sqrt{2}$.

The **asymptotes** of the rectangular hyperbola $xy=c^2$ are $x=0$ and $y=0$.

The **tangent** at the point (x', y') to $xy=c^2$ is
$$xy'+yx'=2c^2$$

DEFINITION OF A PARABOLA

A parabola can be defined as the locus of a point that moves so that its distance from a fixed point, called the focus S, is equal to its distance from a fixed line, the directrix.

In Fig 12.2, if S is the focus and X the foot of the perpendicular from S on to the fixed line, the parabola will be symmetrical about SX. The parabola clearly passes through A, the midpoint of SX, for $SA=AX$, so A has the property common to all points on the parabola. Taking A as origin, and AS produced as the x-axis, if the coordinates of any point on the parabola are (x, y), since $PS=PM$, where M is the foot of the perpendicular from P on to the directrix,

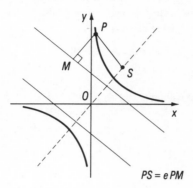

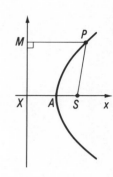

Fig 12.2

$$PS = e\,PM$$

$$PS^2=PM^2$$
i.e.
$$(x-a)^2+y^2=(x+a)^2$$
i.e.
$$y^2=4ax$$

TRANSLATION OF AXES

If the axis of symmetry of the parabola is $y=b$, instead of the x-axis $y=0$, and the vertex is at (c,b) instead of $(0,0)$ (see Fig 12.3), the equation of the parabola is

$$(y-b)^2=4A(x-c)$$

where A is the distance of the vertex from the directrix.

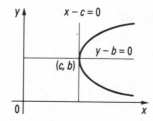

Fig 12.3

TANGENT WITH GIVEN GRADIENT m

Any straight line with gradient m can be written in the form $y=mx+c$, for some c. We want to find the value of c for which this line is a tangent

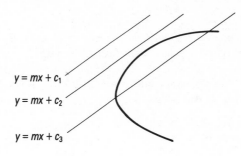

$y=mx+c_1$

$y=mx+c_2$

$y=mx+c_3$

Fig 12.4

to $y^2=4ax$ (Fig 12.4). The line meets the parabola where $(mx+c)^2=4ax$,

i.e. where $\qquad m^2x^2+2x(mc-2a)+c^2=0$

If the straight line is a tangent, the quadratic equation will have two equal roots, i.e.

$$4(mc-2a)^2=4m^2c^2$$

i.e. $\qquad c=a/m$

so the equation of the tangent is

$$y=mx+a/m$$

TANGENT AT THE POINT (x', y')

Differentiating the equation of the parabola to find the gradient,

$$2y\frac{dy}{dx}=4a$$

i.e. $\qquad \dfrac{dy}{dx}=\dfrac{2a}{y}$

so the equation of the tangent at (x', y') is

$$y-y'=\frac{2a}{y'}(x-x')$$

i.e. $\qquad yy'=2a(x+x')$, using $(y')^2=4ax'$ $\hfill$ (1)

NORMAL TO A PARABOLA

The normal at any point is the line through that point perpendicular to the tangent there, so that the normal at (x', y') is

$$y-y' = -\frac{y'}{2a}(x-x')$$

For all except the simplest numerical cases, it is almost invariably easier to use the parametric form (see p145).

Example 12.1 Find the equation of the tangent and normal to the parabola $y^2=12x$ at the point $(3,6)$.

Comparing $y^2=12x$ with $y^2=4ax$, we see $a=3$, so using equation (1), the tangent is

$$6y=6(x+3)$$
i.e. $\qquad y=x+3$

Since the gradient of this tangent is 1, the gradient of the normal at the point is -1, so the equation of the normal is

$$y-6=-1(x-3)$$
i.e. $\qquad x+y=9$

RECTANGULAR HYPERBOLA; DEFINITION

If instead of satisfying $PS=PM$, the point P is such that $PS=ePM$, P describes a general conic; if $e=1$, we have seen the locus is a parabola.

If $\qquad 0<e<1$, the locus is an ellipse.
if $\qquad e=1$, a parabola,
if $\qquad e>1$, a hyperbola, where if
$\qquad e=\sqrt{2}$, we have a special case, a rectangular hyperbola

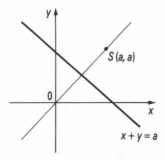

Fig 12.5

Take the coordinates of S as (a, a); the equation of the directrix as $x+y=a$ (Fig 12.5). Then the perpendicular distance of any point (x',y') from the straight line $x+y=a$ is

$$\frac{x'+y'-a}{\sqrt{2}}$$

so that if $PS=\sqrt{2}PM$, i.e. $PS^2=2PM^2$

i.e. $\qquad (x-a)^2+(y-a)^2=2\left(\frac{x+y-a}{\sqrt{2}}\right)^2$

whence $\qquad\qquad xy=\tfrac{1}{2}a^2$

The form $xy=c^2$ or $y=c^2/x$ is more commonly used; in this case the focus is $(c\sqrt{2}, c\sqrt{2})$ and the directrix is $x+y=c\sqrt{2}$.

SYMMETRY OF RECTANGULAR HYPERBOLA

Since the equation $xy=c^2$ is satisfied by $(-x', -y')$ for all (x', y') such that $x'y'=c^2$, the curve is symmetrical about the origin. Corresponding to the one focus we already have, $(c\sqrt{2}, c\sqrt{2})$ there will also be a second focus $(-c\sqrt{2}, -c\sqrt{2})$, and a second directrix $x+y=-c\sqrt{2}$

ASYMPTOTES OF A RECTANGULAR HYPERBOLA

Writing the equation in the form $y=c^2/x$, we see that as x becomes large, c^2/x becomes small, as small as we wish, so that the curve becomes as close as we wish to $y=0$. The straight line $y=0$ is called an asymptote; similarly $x=0$ is also an asymptote.

Translating the axes, to obtain the hyperbola $(x-a)(y-b)=c^2$, we see that can be written

$$y-b=c^2/(x-a)$$

so that as x becomes large, $y-b$ becomes as small as we wish, and one asymptote is $y=b$; similarly, the other asymptote is $x=a$.

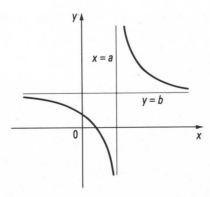

Fig 12.6

EQUATION OF A RECTANGULAR HYPERBOLA, GIVEN THE ASYMPTOTES

Given the asymptotes $x=a$, $y=b$, we can see that we have a family of hyperbolae $(x-a)(y-b)=k$, for differing values of k. More generally, if the equations of the two asymptotes are $a_1x+b_1y+c_1=0$, $a_2x+b_2y+c_2=0$, the family of rectangular hyperbolae with these as asymptotes is

$$(a_1x+b_1y+c_1)(a_2x+b_2y+c_2)=k$$

TANGENT AT (x', y')

Differentiating $xy=c^2$,

$$x\frac{dy}{dx}+y=0$$

$$\frac{dy}{dx}=-\frac{y'}{x'}\text{at } (x', y')$$

so the equation of the tangent is

$$y-y'=-\frac{y'}{x'}(x-x')$$

i.e. $xy'+yx'=2c^2$, using $x'y'=c^2$

EQUATION OF NORMAL

Since the normal is perpendicular to the tangent, the gradient is x'/y' and the equation is

$$xx'-yy'=(x')^2-(y')^2$$

Example 12.2 Find the equation of the tangent and the normal at $(3,12)$ to the curve $xy=36$.

Comparing $xy=36$ with $xy=c^2$, $c=6$, so the equation of the tangent is

$$3y+12x=72$$
i.e. $$y+4x=24$$

and the normal, having gradient $\frac{1}{4}$, is

$$y-12=\frac{1}{4}(x-3)$$
i.e. $$4y=x+45.$$

QUESTIONS

1   Find the equation of the parabola focus (4,0), directrix $x=-4$.
2   Find the equation of the parabola focus (4,3) directrix $x=-2$.
3   Find the value of c if $y=2x+c$ is a tangent to the parabola $y^2=6x$.
4   Find the equation of the tangent to the parabola $y^2=8x$ at the point (2,4). Find also the equation of the normal at this point.
5   Find an equation satisfied by l, m, n and a if $lx+my+n=0$ is a tangent to $y^2=4ax$.
6   The tangent at the point (x',y') meets the asymptotes $y=0$, $x=0$ at P_1,P_2 respectively. Prove that the area of the triangle OP_1P_2 is independent of x' and y'.
7   With the data of Q.6 prove that (x',y') is always the midpoint of P_1,P_2.
8   The tangent at $P_1(x_1y_1)$ meets an asymptote at Q_1; the tangent at P_2 (x_2, y_2) meets that asymptote at Q_2; show that P_1P_2 always passes through the midpoint of Q_1,Q_2.
9   The tangent at $P(x',y')$ meets the asymptotes at T_1,T_2: the normal at P meets the asymptotes at N_1,N_2,N_1 being on OT_1. Prove that $OT_1:OT_2=ON_2:N_1O$.
10   Find the relation satisfied by l, m, n, and c if $lx+my+n=0$ touches $xy=c^2$.

ELLIPSE AND HYPERBOLA

CONTENTS

The **eccentricity** of an ellipse is less than 1, i.e. $e>1$.

If the equation of an ellipse is

$$\frac{x^2}{a^2}+\frac{y^2}{b^2}=1,$$

the **foci** are $(ae,0)$ and $(-ae,0)$; the **directrices** are $x=a/e$, $x=-a/e$

The **tangent at the point** (x',y') is

$$\frac{xx'}{a^2}+\frac{yy'}{b^2}=1,$$

the **tangents with given gradient** m are

$$y=mx\pm\sqrt{(a^2m^2+b^2)}$$

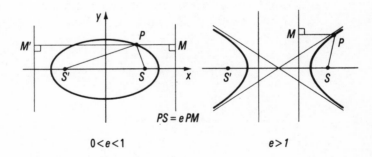

$$PS = e\,PM$$

Fig 13.1 $0<e<1$ $e>1$

The **eccentricity** of a hyperbola is greater than 1, i.e. $e>1$.

If the equation of a hyperbola is

$$\frac{x^2}{a^2}-\frac{y^2}{b^2}=1$$

the **foci** are also at $(ae,0)$ and $(-ae,0)$; the **directrices** are $x=a/e$, $x=-a/e$

The **tangent** at (x',y') is

$$\frac{xx'}{a^2}-\frac{yy'}{b^2}=1,$$

the **tangents with given gradient** m are

$$y=mx\pm\sqrt{(a^2m^2-b^2)}$$

The **asymptotes** of $\dfrac{x^2}{a^2}-\dfrac{y^2}{b^2}=1$ are

$$\frac{x^2}{a^2}-\frac{y^2}{b^2}=0, \text{ i.e. } \frac{x}{a}-\frac{y}{b}=0 \text{ and } \frac{x}{a}+\frac{y}{b}=0$$

ELLIPSE AND HYPERBOLA

In Chapter 12, we saw that the locus of a point P whose distance from a fixed point S, the focus, is proportional to its distance from a fixed line, the directrix, is called an ellipse if the constant of proportion e is such that $0<e<1$; a hyperbola if $e>1$, with the special case $e=\sqrt{2}$ giving a form called a rectangular hyperbola, the asymptotes of which are at right angles. The ellipse and hyperbola clearly have much in common, and we shall see many similarities between the corresponding equations.

EQUATION OF AN ELLIPSE

The simplest form for the equation of an ellipse occurs when we take one focus at $(ae,0)$, and the corresponding directrix as $x=a/e$. Then $PS^2=(x-ae)^2+y^2$, $PM=(a/e-x)$, so that since $PS^2=e^2PM^2$,

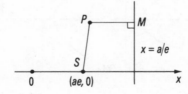

Fig 13.2

$$(x-ae)^2+y^2=e^2(a/e-x)^2$$

i.e. $\qquad x^2(1-e^2)+y^2=a^2(1-e^2) \qquad\qquad (1)$

Writing $b^2=a^2(1-e^2)$, and dividing by b^2, we have

$$\frac{x^2}{a^2}+\frac{y^2}{b^2}=1$$

EQUATION OF HYPERBOLA

To find the equation of a hyperbola, we proceed exactly as above, until we have obtained equation (1). Now, since $e>1$ we cannot write $b^2=a^2(1-e^2)$, we have to write $b^2=a^2(e^2-1)$, and the equation of the hyperbola is

$$\frac{x^2}{a^2}-\frac{y^2}{b^2}=1$$

SYMMETRY OF AN ELLIPSE

Since the equation of an ellipse only contains even powers of x, if (x',y') lies on this ellipse, so does $(-x',y')$, and the curve is symmetrical about the y-axis; similarly the curve is symmetrical about the x-axis, as it only contains even powers of y. Thus there will be a second focus S' at $(-ae,0)$ and a second directrix $x=-a/e$.

SUM OF FOCAL DISTANCES OF A POINT ON AN ELLIPSE

We may be familiar with the rough method used to draw an ellipse, with a length of string pinned at each end to a board, allowing a pencil to move along the string. This method uses the property that $PS+PS'$ is constant.

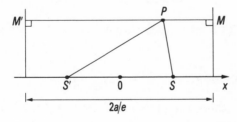

Fig 13.3

Since $PS=ePM$, and $PS'=ePM$
$$PS+PS'=e(PM+PM')$$
$$=e(2a/e)$$

Since the distance between the two directrices is constant, for all positions of P,

$$PS+PS'=2a$$

TANGENT AT (x′, y′)

Differentiating the equation of the ellipse

$$\frac{2x}{a^2}+\left(\frac{2y}{b^2}\right)\frac{dy}{dx}=0$$

so the gradient at (x',y') is $-b^2x'/a^2y'$, and the equation of the tangent is

$$y-y'=-\left(\frac{b^2x'}{a^2y'}\right)(x-x')$$

which reduces to $\dfrac{xx'}{a^2}+\dfrac{yy'}{b'}=1$

after using $\dfrac{(x')^2}{a^2}+\dfrac{(y')^2}{b^2}=1$

EQUATION OF THE NORMAL AT (x′, y′)

Since the gradient of the normal is $(a^2y')/(b^2x')$, the equation is

$$y-y'=\frac{a^2y'}{b^2x'}(x-x')$$

which reduces to $a^2xy'-b^2yx'=(a^2-b^2)x'y'$

This form should not be committed to memory, but rather the method should be used in any numerical examples.

Example 13.1 Find the equation of the tangent and normal at $(1,1)$ to the ellipse $2x^2+y^2=3$.

Differentiating, $4x+2y\dfrac{dy}{dx}=0$

so the gradient at $(1,1)$ is -2, and the equation of the tangent is

$$y-1=-2(x-1)$$
i.e. $2x+y=3$

This could have been obtained by rearranging the equation of the ellipse

$$\frac{x^2}{(\frac{3}{2})}+\frac{y^2}{3}=1$$

and using $\dfrac{xx'}{a^2}+\dfrac{yy'}{b^2}=1$

to give
$$\frac{1x}{(\frac{3}{2})}+\frac{1y}{3}=1$$

i.e.
$$2x+y=3$$

To find the normal, since this is perpendicular to the tangent the gradient is $\frac{1}{2}$, so the equation is

$$y-1=\tfrac{1}{2}(x-1)$$
i.e.
$$2y=x+1$$

EQUATION OF TANGENT WITH GRADIENT m TO AN ELLIPSE

The equation of the tangent will be of the form $y=mx+c$, for some c. This meets the ellipse where

$$\frac{x^2}{a^2}+\frac{(mx+c)^2}{b^2}=1$$

i.e.
$$(b^2+a^2m^2)x^2+2a^2mcx+a^2(c^2-b^2)=0$$

If this line is to be a tangent, the quadratic equation must have equal roots, and

$$4a^4m^2c^2=4(b^2+a^2m^2)a^2(c^2-b^2)$$
whence
$$c^2=a^2m^2+b^2$$

so the equation of a tangent with gradient m is

$$y=mx+\sqrt{(a^2m^2+b^2)}$$

There are, of course, two tangents with given gradient m,

$$y=mx\pm\sqrt{(a^2m^2+b^2)} \qquad\qquad (2)$$
since
$$c=\pm\sqrt{(a^2m^2+b^2)}$$

Example 13.2 Find the equation of the tangent with gradient 2 to the ellipse $x^2/4+y^2/9=1$

Any straight line with gradient 2 has equation of the form $y=2x+c$. This will be a tangent to the ellipse if

$$\frac{x^2}{4}+\frac{(2x+c)^2}{9}=1$$

has equal roots, i.e. if

$$25x^2+16cx+4(c^2-9)=0$$

has equal roots. Thus

$$(16c)^2=4\times25\times4(c^2-9)$$
$$16c^2=25c^2-225$$
$$c^2=25, c=\pm5$$

the two tangents with gradient 2 are $y=2x\pm5$.

Alternatively, comparing with the standard form of the equation of the ellipse, we have $a^2=4$, $b^2=9$, and $m=2$, so using equation (2), the equation of the tangents is

$$y=2x\pm\sqrt{(4\times2^2+9)}$$

i.e. $y=2x\pm5$

HYPERBOLA, FOCI, DIRECTRICES AND ASYMPTOTES

We have seen that the equation of a hyperbola can be written in the form

$$\frac{x^2}{a^2}-\frac{y^2}{b^2}=1 \tag{3}$$

where $b^2=a^2(e^2-1)$, the foci are $(\pm ae,0)$, and the directrices are $x=\pm a/e$. If we are given the equation in the form (3), then we can deduce the foci, using $b^2=a^2(e^2-1)$. The equation of the hyperbola can be written

$$\left(\frac{x}{a}-\frac{y}{b}\right)\left(\frac{x}{a}+\frac{y}{b}\right)=1$$

so that

$$\frac{x}{a}-\frac{y}{b}=\frac{1}{\frac{x}{a}+\frac{y}{b}}$$

or

$$\frac{x}{a}+\frac{y}{b}=\frac{1}{\frac{x}{a}-\frac{y}{b}}$$

Thus if x and y are each large and of the same sign, $1/(x/a+y/b)$ is as small as we wish, so that $x/a-y/b=0$ is an asymptote. Similarly $x/a+y/b=0$ is the other asymptote.

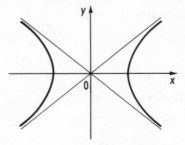

Fig 13.4

As with the ellipse, the hyperbola is symmetrical about the x-axis and the y-axis.

EQUATION OF TANGENTS AND NORMAL

We can show in exactly the same manner as for the ellipse that the equation of the tangent at the point (x',y') is

$$\frac{xx'}{a^2}-\frac{yy'}{b^2}=1 \tag{4}$$

and the equation of the tangents with a given gradient m are

$$y=mx\pm\sqrt{(a^2m^2-b^2)} \tag{5}$$

The equation of the normal at (x',y') is

$$a^2y'x+b^2x'y=(a^2+b^2)x'y'$$

Example 13.3 Find the foci, directrices and asymptotes of the hyperbola $9x^2-25y^2=225$, and the equation of the tangent at $(\frac{25}{3},4)$.

Rearranging the equation of the hyperbola,

$$\frac{x^2}{25}-\frac{y^2}{9}=1$$

and comparing with (3), we see $a^2=25$, $b^2=9$, so that

$$25=9(e^2-1)$$
and $\qquad e=\tfrac{1}{3}\sqrt{34}.$

Thus the foci are $(\pm\tfrac{5}{3}\sqrt{34}, 0)$ and the directrices are $x=\pm15/\sqrt{34}$, and the asymptotes

$$\frac{x}{25}-\frac{y^2}{9}=0$$

i.e. $\qquad \frac{x}{5}-\frac{y}{3}=0$ and $\frac{x}{5}+\frac{y}{3}=0$

To find the tangent at $(25/3,4)$, we can differentiate (3) to find the gradient , or substituting in (4), we have

$$\frac{x}{25}\left(\frac{25}{3}\right)-\frac{y}{9}(4)=1$$

i.e. $\qquad 3x-4y=9$

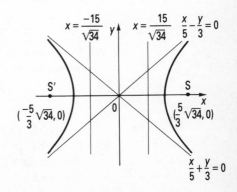

Fig 13.5

QUESTIONS

1. Find the coordinates of the foci and the eccentricity of
    $$4x^2+9y^2=36$$
    and of $\qquad\qquad 9x^2+4y^2=36$
    Draw sketches illustrating the two elipses.

2. Find the equation of the tangent and of the normal at (2,1) to the ellipse $2x^2+3y^2=11$.
    Deduce the equation of the tangent and normal at $(2, -1)$.

3. Find the equations of the tangents with gradient 4 to the ellipse

    $$\frac{x^2}{9}+\frac{y^2}{25}=1$$

4. Find the equations of the axes of symmetry, the coordinates of the foci, and the equations of the directrices of the ellipse

    $$\frac{(x+2)^2}{16}+\frac{(y-1)^2}{7}=1$$

5. Find the points of contact of the tangents of gradient 1 with the ellipse

    $$\frac{x^2}{64}+\frac{y^2}{225}=1$$

    and obtain the equations of the two normals to this ellipse which have gradient -1.

6. Find the foci, directrices and asymptotes of

    $$\frac{x^2}{4}-\frac{y^2}{9}=1$$

    and $\qquad \dfrac{x^2}{9}-\dfrac{y^2}{4}=1$

    and illustrate the two hyperbolae by sketches.

7. Find the equation of the tangent and the normal to

    $$2x^2-y^2=1$$

    at the point (1,1).

8. Find the equations of the tangents with gradient 3 to the hyperbola

    $$\frac{x^2}{25}-\frac{y^2}{64}=1$$

9. Find the equations of the axes of symmetry, the foci, directrices and asymptotes of

    $$\frac{(x-1)^2}{16}-\frac{(y+2)^2}{25}=1$$

    illustrating these results by a sketch.

10. Find the points of contact of the tangents with gradient 1 to the hyperbola $x^2/25-y^2/16=1$, and deduce the normals to the curve with gradient -1.

PARAMETRIC FORMS

CONTENTS

NOTES

Any point on the **circle** $x^2+y^2=r^2$ can be taken in the form $(r\cos\theta, r\sin\theta)$,
any point on the **parabola** $y^2=4ax$ as $(at^2,2at)$,
any point on the **rectangular hyperbola** $xy=c^2$ as $(ct,c/t)$,
any point on the **ellipse** $x^2/a^2+y^2/b^2=1$ as $(a\cos\theta, b\sin\theta)$,
any point on the **semi-cubical parabola** $y^2=x^3$ as (t^2,t^3).

When these parameters are used,
the **tangent** at $(r\cos\theta, r\sin\theta)$ is

$$x\cos\theta+y\sin\theta=r$$

the **tangent** at $(at^2,2at)$ is

$$ty=x+at^2$$

the **normal** at $(at^2,2at)$ is

$$y+tx=2at+at^3$$

the **tangent** at $(ct,c/t)$ is

$$x+t^2y=2ct$$

CIRCLE

We can see in Fig 14.1 that the point P $(r\cos\theta, r\sin\theta)$ lies on the circle centre $(0,0)$ radius r. When $\theta=0$, the point is at A, $(r,0)$. As θ increases, the point P moves anti-clockwise around the circle, reaching B when $\theta=\pi/2$, and so on as θ increases. Any value of θ gives one point on the curve; any point on the curve corresponds to one value of θ between 0 and 2π. θ is called a *parameter*, and many properties of curves can often be investigated easily if we use a parametric form.

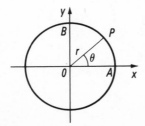

Fig 14.1

EQUATION OF THE TANGENT AT
$(r\cos\theta, r\sin\theta)$

Any point on the circle $x^2+y^2=r^2$ can be written in this form. To find the equation of the tangent at the point $(r\cos\theta, r\sin\theta)$, we can substitute in the form $xx'+yy'=r^2$ (from p136), to obtain,

$$xr\cos\theta+yr\sin\theta=r^2$$
i.e. $$x\cos\theta+y\sin\theta=r^2$$

or we can differentiate the parametric form. To find dy/dx, we use

$$\frac{dy}{dx}=\frac{dy/d\theta}{dx/d\theta}$$

Now since $x=r\cos\theta$, $dx/d\theta=-r\sin\theta$,
and since $y=r\sin\theta$, $dy/d\theta=r\cos\theta$, so that

$$\frac{dy}{dx}=-\frac{r\cos\theta}{r\sin\theta}=-\cot\theta$$

and the equation of the tangent at $(r\cos\theta, r\sin\theta)$ is

$$y-r\sin\theta=-\cot\theta(x-r\cos\theta)$$
i.e. $$x\cos\theta+y\sin\theta=1,\text{ using }\cos^2\theta+\sin^2\theta=1$$

PARABOLA

Consider the parabola $y^2=4ax$. The points $(\frac{1}{4}a,a)$, $(a,2a)$, $(4a,4a)$ all lie on the parabola, and any point of the form $(at^2,2at)$ lies on the parabola. For the three points given as examples, we see that $t=\frac{1}{2}$, 1, and 2 respectively.

TANGENT AT $(at^2,2at)$

As with the circle, we could recall the equation of the tangent from p125, $yy'=2a(x+x')$, and substituting $x'=at^2$, $y'=2at$, we have

$$2aty=2a(x+at^2)$$
i.e. $$ty=x+at^2$$

Alternatively, we can use $\frac{dy}{dx}=\frac{dy/dt}{dx/dt}$, where $\frac{dy}{dt}=2a$, $\frac{dx}{dt}=2at$, so that the equation of the tangent at $(at^2,2at)$ is

$$y-2at=\frac{1}{t}(x-at^2)$$
i.e. $$ty=x+at^2$$

Example 14.1 Show that any two perpendicular tangents to the parabola $y^2=4ax$ intersect on the directrix.

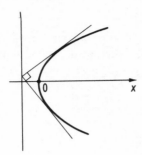

Fig 14.2

If the points of contact are $(at^2, 2at)$, $(am^2, 2am)$, the equation of the tangents are $ty=x+at$ and $my=x+am$. If these are perpendicular, $(1/t)(1/m)=-1$, i.e. $mt=-1$. Now the tangents meet where

$$ty=x+at^2 \qquad\qquad (1)$$
$$\text{and} \quad my=x+am^2 \qquad\qquad (2)$$

If the point of intersection lies on the directrix, then the x coordinate must be $-a$. Multiplying (1) by m and (2) by t and subtracting,

$$0=x(m-t)+amt(t-m)$$
i.e. $\qquad 0=x+a$, using $mt=-1$
i.e. $\qquad x=-a$ for all t, m such that $mt=-1$.

EQUATION OF NORMAL AT $(at^2, 2at)$

Since the gradient of the tangent is $1/t$, the gradient of the normal is $-t$, and the equation of the normal is

$$y-2at=-t(x-at^2)$$
i.e. $\qquad y+tx=2at+at^3 \qquad\qquad (1)$

Example 14.2 The normal to $y^2=4ax$ at $(at^2, 2at)$ meets the coordinate axes in G and H. Find the locus of the midpoint of GH as t varies.

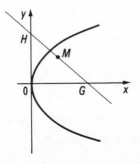

Fig 14.3

If G is the point at which the normal meets $y=0$, from equation (1) we see that at G, $x=a(2+t^2)$. At H, where $x=0$, $y=2at+at^3$, i.e. $y=at(2+t^2)$. The coordinates of the midpoint M of GH are therefore

$$x=\tfrac{1}{2}a(2+t^2),\ y=\tfrac{1}{2}at(2+t^2)$$

These would serve as the parametric form for the locus, but the cartesian form can be obtained easily by noticing that $at^2=2x-2a$,

so that
$$y^2=\tfrac{1}{4}a^2t^2(2+t^2)^2$$

i.e.
$$y^2=\tfrac{1}{4}a(2x-2a)\frac{4x^2}{a^2}$$

i.e.
$$ay^2=2x^2(x-a)$$

EQUATION OF CHORD THROUGH $(ap^2, 2ap)$, $(aq^2, 2aq)$

The equation of the straight line through the points $(ap^2,2ap)$, $(aq^2,2aq)$ is

$$\frac{y-2ap}{2aq-2ap}=\frac{x-ap^2}{aq^2-ap^2}$$

i.e.
$$y-2ap=\frac{2a(q-p)}{a(q-p)(q+p)}(x-ap^2)$$

$$a(q+p)(y-2ap)=2(x-ap^2)$$
$$(p+q)y=2x+2apq$$

Notice first, that this equation is symmetrical in p and q, secondly, that as q tends to p, this equation becomes that of the tangent at $(ap^2,2ap)$.

Example 14.3 The chord joining points P and Q on a parabola always passes through the focus. Show that the tangents at P and Q are always perpendicular.

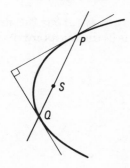

Fig 14.4

Since the chord passes through $(a,0)$,

$$(p+q)\times0=2a+2apq,\ \text{i.e. } pq=-1$$

But we have seen that the gradient of the tangent at P is $1/p$, at Q is

$1/q$, so that $pq=-1 \Rightarrow (1/p)(1/q)=-1$, the tangents at P and Q are always perpendicular.

THE RECTANGULAR HYPERBOLA $x=ct, y=c/t$

All points on the curve $xy=c^2$ can be described by the parametric form $x=ct$, $y=c/t$. When t is positive, P lies in the first quadrant; when t is negative, P lies in the third quadrant. We see that when t is small and positive, P lies close to the y-axis (since ct is small); the positive vertex A corresponds to $t=1$, and when t is large and positive, the point P is close to the x-axis.

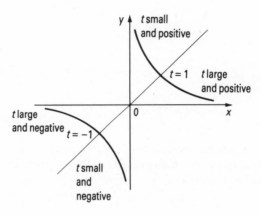

Fig 14.5

TANGENT AT $(ct, c/t)$

Since $x=ct$, $dx/dt=c$; since $y=c/t$, $dy/dt=-c/t^2$,

$$\frac{dy}{dx}=\frac{-c/t^2}{c}=-\frac{1}{t^2}$$

and the equation of the tangent is

$$y-\frac{c}{t}=\left(-\frac{1}{t^2}\right)(x-ct)$$

i.e. $t^2y+x=2ct$

This could also have been obtained by putting $x=ct$, $y=c/t$ in the equation $xy'+yx'=2c^2$.

EQUATION OF THE NORMAL

The gradient of the normal is t^2, so the equation of the normal is

$$y-c/t=t^2(x-ct)$$
i.e. $ty-t^3x=c(1-t^4)$

Example 14.4 P is any point on a rectangular hyperbola. $x=ct$, $y=c/t$. The normal at P meets the hyperbola again at Q. If M is the midpoint of PQ, show that PO is perpendicular to OM for all positions of P.

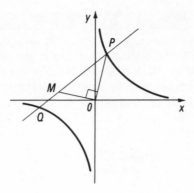

Fig 14.6

The normal at P is $ty-t^3x=c(1-t^4)$. Take Q as $(cq,c/q)$. This lies on the normal at P if

$$ct/q-ct^3q=c(1-t^4)$$

i.e. $$t^3q^2+(1-t^4)q-t=0$$

We know that one root of this equation is $q=t$, since one of the two points in which this line meets the hyperbola is P, so that one factor must be $(q-t)$, thus

$$(q-t)(t^3q+1)=0$$

$$q=t \text{ or } -\frac{1}{t^3}$$

so the coordinates of Q are $(-c/t^3, -ct^3)$.

Now the coordinates of M are

$$\frac{1}{2}c\left(t-\frac{1}{t^3}\right), \frac{1}{2}c\left(\frac{1}{t}-t^3\right)$$

so the gradient of OM is

$$\frac{\frac{1}{2}c\left(\frac{1}{t}-t^3\right)}{\frac{1}{2}c\left(t-\frac{1}{t^3}\right)}$$

$$=\frac{t^2(1-t^4)}{t^4-1}=-t^2$$

The gradient of OP is $\dfrac{c/t}{ct}=\dfrac{1}{t^2}$, and since

$$\left(-t^2\right)\left(\frac{1}{t^2}\right)=-1$$

OM is perpendicular to *OP* for all positions of *P*.

Notice that when solving the quadratic for *q*, we recalled that we knew one value of *q*, so that we knew one factor, and therefore could find the other easily.

THE ELLIPSE $x=a \cos \theta, y=b \sin \theta$

We see that $x=\cos \theta, y=b \sin \theta$ satisfies

$$\frac{x^2}{a^2}+\frac{y^2}{b^2}=1$$

for all values of θ, since

$$\frac{(a \cos \theta)^2}{a^2}+\frac{(b \sin \theta)^2}{b^2}=1$$

for all θ. Unlike most other parameters, though, θ has a geometric interpretation, and Fig 14.7 shows a circle, centre *O*, radius *a* (called the auxiliary circle) and the relation of θ to the point *P'* on that circle with the same *x* coordinate as *P*.

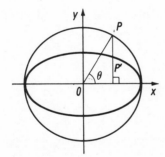

Fig 14.7

EQUATION OF THE TANGENT AND NORMAL AT $a \cos \theta, b \sin \theta$

Using the form on p136 and substituting $x=a \cos \theta, y=b \sin \theta$, the equation of the tangent is

$$\frac{x \cos \theta}{a}+\frac{y \sin \theta}{b}=1$$

The equation of the normal is found to be

$$ax \sin \theta-by \cos \theta=(a^2-b^2) \sin \theta \cos \theta$$

Example 14.5 The tangent at a variable point P meets the axes at Q and R. S is the fourth vertex of the rectangle $OQSR$. Show that as P varies, the locus of Q is $a^2y^2+b^2x^2=x^2y^2$.

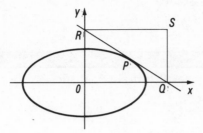

Fig 14.8

The equation of the tangent at P is

$$\frac{x \cos \theta}{a}+\frac{y \sin \theta}{b}=1$$

this meets $y=0$ at $(a/\cos \theta, 0)$ and $x=0$ at $(0, b/\sin \theta)$ so the coordinates (X, Y) of S are

$$X=\frac{a}{\cos \theta}, \ Y=\frac{b}{\sin \theta}$$

To find the locus of S, eliminate $\cos \theta$ and $\sin \theta$, using $\cos^2\theta+\sin^2\theta=1$,

i.e. $$\left(\frac{a}{X}\right)^2+\left(\frac{b}{Y}\right)^2=1$$

the locus of S is $a^2y^2+b^2x^2=x^2y^2$.

THE SEMI-CUBICAL PARABOLA $x=t^2$, $y=t^3$

All points on the curve $y^2=x^3$ can be described parametically by $x=t^2$, $y=t^3$. This curve is symmetrical about the x-axis, and only exists for positive values of x.

Example 14.6 Find the equation of the tangent at the point (t^2, t^3) on the curve $x^3=y^2$, and the coordinates of the other point at which the tangent meets the curve again (Fig 14.9).

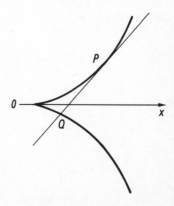

Fig 14.9

We have to use

$$\frac{dy}{dx}=\frac{\frac{dy}{dt}}{\frac{dx}{dt}}$$

$dy/dt=3t^2$, $dx/dt=2t$, so $dy/dx=\frac{3}{2}t$, and the equation of the tangent is

$$y-t^3=\frac{3}{2}t(x-t^2)$$

i.e. $\qquad 2y-3tx+t^3=0$

If this meets the curve again at Q, take Q as (q^2, q^3), and since Q lies on the tangent at P,

$$2q^3-3tq^2+t^3=0$$

Since the tangent touches the curve at P, there are two roots $q-t$, so that $(q-t)^2$ is a factor of $2q^3-3tq^2+t^3$, and we can find the other linear factor easily by considering the coefficient of q^3, and the constant term, thus

$$2q^3-3tq^2+t^3=0$$
$$\Rightarrow \qquad (q-t)^2(2q+t)=0$$
$$\Rightarrow \qquad\qquad q=t \text{ (twice) or } -\tfrac{1}{2}t$$

the coordinates of Q are $(-\tfrac{1}{2}t)^2, (-\tfrac{1}{2}t)^3$, i.e. $(\tfrac{1}{4}t^2, -\tfrac{1}{8}t^3)$.

QUESTIONS

1 Find suitable parametric forms for the following:
   (a) $x^2+y^2=25$ (b) $x^2+y^2=\tfrac{1}{4}$
   (c) $y^2=4x$ (d) $y^2=-36x$ (e) $x^2=y$

2 Points P and Q on the circle $x^2+y^2=r^2$ have parameters θ and φ respectively. Prove that the equation of the chord PQ can be written

$$x\cos\tfrac{1}{2}(\theta+\varphi)+y\sin\tfrac{1}{2}(\theta+\varphi)=r\cos\tfrac{1}{2}(\theta-\varphi).$$

3 Points P and Q on the parabola $y^2=4ax$ have parameters p and q. Find the equation of the chord PQ, and show that if PQ passes through the focus, $pq=-1$.

4 The tangent at a point P on $y^2=4ax$ meets the y-axis at Y; the normal to the curve at P meets the x-axis at X. Show that the locus of the midpoint of XY is $2y^2=a(x-a)$.

5 The normal at a point P on $y^2=4ax$ meets the x-axis Ox at N. Show that $2a<PN^2/ON<4a$, whatever the position of P.

6 Write down parametric forms for the following:
   (a) $xy=9$ (b) $4xy=25$ (c) $xy=-c^2$
   (d) $9xy=-4$ (e) $(x-1)(y+2)=c^2$

7   The tangent at a variable point P $(ct, c/t)$ on $xy=c^2$ meets the asymptotes at T_1, T_2. Find the coordinates of T_1 and T_2 in terms of t and show that the area of the triangle OT_1T_2 is constant for all positions of P.

8   With the data of Q.2 show that P is always the midpoint of T_1T_2.

9   The normal at a variable point P on $xy=c^2$ meets $y=x$ at Q. Show that $PQ=PO$, where O is the origin.

10   The tangent at a variable point P meets the x-axis at T_1, the y-axis at T_2; the normal at P meets the x-axis at N_1, the y-axis at N_2. Show that $OT_1 \times ON_1 = OT_2 \times ON_2$.

11   Find parametric forms for the following:
     (a) $x^2/9 + y^2/4 = 1$ (b) $25x^2 + 4y^2 = 100$
     (c) $2x^2 + 3y^2 = 1$ (d) $y^2 = x^5$ (e) $y = x^2(x+1)$

12   An ellipse is given by the parametric form $x = 2\cos\theta$, $y = 3\sin\theta$. Find the equation of the tangent to the ellipse at the point parameter θ.

     If this tangent passes through the point $(2,1)$, show that θ satisfies an equation of the form $l\cos\theta + m\sin\theta = n$. Solve this equation to find the two points P and Q at which the tangent passes through $(2,1)$.

13   The tangent at a variable point $P(a\cos\theta, b\sin\theta)$ meets the tangent at $A(a,0)$ at T, and the tangent at $A'(-a,0)$ at T'. Show that $AT \times A'T'$ is constant for all positions of P.

14   Given that the tangent at P to the curve $x=t^2$, $y=t^3$ passes through the point $(1,0)$, find the two possible parameters for P, given that P is not the origin.

15   Find the equation of the tangent at P, the point $t=\sqrt{2}$, to the curve $x=3t^2$, $y=2t^3$. Find also the parameter of the point Q at which this tangent meets the curve again, and show that the tangent at P is the normal at Q.

POLAR COORDINATES

CONTENTS

NOTES

To change cartesian coordinates to polars,

$$x=r\cos\theta, y=r\sin\theta$$

To change polar coordinates to cartesians,

$$r^2=x^2+y^2, \tan\theta=\frac{y}{x}$$

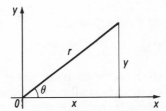

Fig 15.1

The area of a sector bounded by the curve $r=f(\theta)$, and the radii $\theta=\alpha$ and $\theta=\beta$ is $\frac{1}{2}\int_{\alpha}^{\beta}r^2 d\theta$

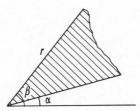

Fig 15.2

POLAR COORDINATES

Cartesian coordinates determine the position of a point relative to two fixed straight lines, the coordinate axes; polar coordinates determine the position of a point P by its distance from a fixed point O (the pole), and the angle made by the line joining P to the pole with a fixed line. Although it is easy to convert from cartesian coordinates to polars, using

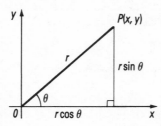

Fig 15.3

$$x = r\cos\theta, \quad y = r\sin\theta$$

and from polars to cartesians, using

$$r = \sqrt{(x^2 + y^2)}, \quad \tan\theta = \frac{y}{x}$$

it is usually better to think in terms of polar coordinates, to realize that the equation $r = a$ describes all points a distance a from the pole, so is the equation of a circle, radius a, centre O; that $\theta = \alpha$ describes all points P such that OP makes a fixed angle α with the base line, and so describes a straight line inclined to the base line at an angle α

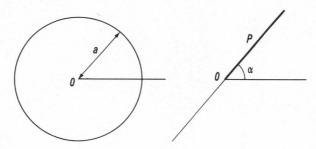

Fig 15.4

THE CURVE $r = a(1 + \cos\theta)$

Given the equation of a curve in the form $r = a(1 + \cos\theta)$, we can see that any one value of α determines one value of r, so that for any one given angle made by OP with the base line, we can find the distance of P from O for that angle. Thus when $\theta = 0$, $\cos\theta = 1$ and $r = 2a$; as θ increases, $\cos\theta$ decreases and so r decreases, taking the value a when $\theta = \pi/2$, and the value 0 when $\theta = \pi$; r then increases again, being equal to a when $\theta = 3\pi/2$, finally $r = 2a$ when $\theta = 2\pi$. We have the curve sketched in Fig 15.5.

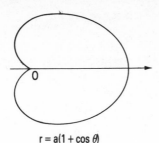

Fig 15.5 $r = a(1 + \cos\theta)$

THE CURVE $r^2=a^2\cos 2\theta$ Following this procedure for $r^2=a^2\cos 2\theta$, when $\theta=0$ we have $r=a$ (strictly $\pm a$, but we usually adopt the convention that r is positive). As θ increases, r decreases, until $r=0$ when $\theta=\pi/4$. When $\pi/4<\theta<3\pi/4$, $\cos 2\theta$ is negative, so there are no real values of r in the interval $\pi/4<\theta<3\pi/4$. As θ increases from $3\pi/4$, r increases, until when $\theta=\pi$, $r=a$. Continuing the curve for values of θ between π and 2π, we obtain the curve in Fig 15.6.

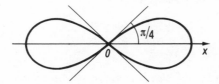

Fig 15.6

THE STRAIGHT LINE $r =p$ cosec $(\alpha-\theta)$ Some polar equations use properties of the curve they describe that require some trigonometry. Consider the straight line a perpendicular distance p from the origin, inclined at an angle α to the base line. Then all points P on this line are such that their distance r from the origin is given by $p=r\sin(\alpha-\theta)$, so the equation of this straight line is $r=p/\sin(\alpha-\theta)$, or $r=p$ cosec $(\alpha-\theta)$.

Expanding the equation $p=r\sin(\alpha-\theta)$, we have

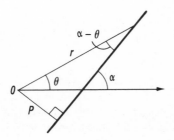

Fig 15.7

$$p=r\sin\alpha\cos\theta-r\cos\alpha\sin\theta$$

Writing $x = r\cos\theta$, $y = r\sin\theta$, we obtain

$$p = x\sin\alpha - y\cos\alpha$$

which we recognize as a straight line in cartesian form.

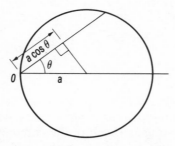

Fig 15.8

THE CIRCLE $r = 2a\cos\theta$

If P is any point on the circle centre $(a,0)$, radius a, from Fig 15.8 we see that $\frac{1}{2}r = a\cos\theta$, so the equation is $r = 2a\cos\theta$. If we wish to express this in cartesian form,

$$r = 2a\cos\theta \Rightarrow r^2 = 2ar\cos\theta$$
$$\Rightarrow x^2 + y^2 = 2ax$$

i.e. $\qquad x^2 + y^2 - 2ax = 0$

a circle centre $(a,0)$, radius a. Notice that since we know $x = r\cos\theta$ and $y = r\sin\theta$, it is often helpful to multiply by r (or r^2) so that we can substitute for $r\cos\theta$ and $r\sin\theta$ rather than $\cos\theta$ and $\sin\theta$

THE SPIRALS $r = a\theta$, $r = ae^{k\theta}$

The equation $r = a\theta$ tells us that as θ increases, so does r, and that any constant increase of θ, say 2π, produces a constant increase in r, so that the curve is a spiral, of constant width $2\pi a$ between points along any one straight line $\theta = \alpha$.

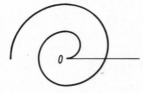

Fig 15.9

By contrast, the equation $r = ae^{k\theta}$ tells us that any increase in θ, say of 2π multiplies r by a factor that often is very large, since

$$r = ae^{k(2\pi + \alpha)} = ae^{2k\pi}e^{k\alpha}$$
$$= (e^{2k\pi})(ae^{k\alpha})$$

and unless k is very small, $e^{2k\pi}$ will be large. Thus we now have a spiral of rapidly increasing width, as in Fig 15.10.

Fig 15.10

AREA OF A SECTOR

To find the area of a sector bounded by the curve and two radii $\theta=\alpha$ and $\theta=\beta$, a small increase $\delta\theta$ in θ produces a corresponding increase δA in the area A, where $\delta A \simeq \frac{1}{2}r^2\delta\theta$. It can be shown that $dA/d\theta=\frac{1}{2}r^2$, so

$$A=\int_{\alpha}^{\beta} \tfrac{1}{2}r^2 d\theta$$

Example 15.1 Find the area enclosed by the curve $r=a(1+\cos\theta)$.

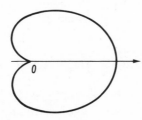

Fig 15.11

This curve, a cardioid, we sketched on p157. The area A enclosed by the curve is

$$\int_0^{2\pi} \tfrac{1}{2}a^2(1+\cos\theta)^2 d\theta$$
$$=\int_0^{2\pi} \tfrac{1}{2}a^2[1+2\cos\theta+\tfrac{1}{2}(1+\cos 2\theta)]d\theta$$

using $\cos^2\theta=\frac{1}{2}(1+\cos 2\theta)$,

$$A=\tfrac{1}{2}a^2\int_0^{2\pi}[\tfrac{3}{2}\theta+2\cos\theta+\tfrac{1}{2}\cos 2\theta]d\theta$$
$$=\tfrac{1}{2}a^2[\tfrac{3}{2}\theta+2\sin\theta=\tfrac{1}{4}\sin 2\theta]_0^{2\pi}$$
$$=\tfrac{1}{2}a^2\times\tfrac{3}{2}\times 2\pi$$
$$=\tfrac{3}{2}\pi a^2$$

Check: It is always important to check solutions where this can be done easily. Here we see that the region enclosed is considerably less than that enclosed by a circle radius $2a$, but greater than that enclosed by a circle radius a.

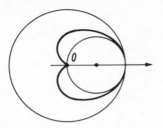

Fig 15.12

1   Sketch the following curves, by using the property given in the equation of the curve:

   (a) $r=2$ (b) $\theta=\pi/2,\ r>0$

   (c) $r=3,\ -\pi/4<\theta<3\pi/4$ (d) $\theta=\pi/3,\ 1<r<2$

2   Sketch the following curves, by finding the values for r corresponding to values of θ between 0 and 2π:

   (a) $r=1+\sin\theta$ (b) $r=2+\sin\theta$

   (c) $r=1+2\sin\theta$ (d) $r=\sin 2\theta$

   (e) $r^2=2/\sin 2\theta$

3   Sketch the following curves:

   (a) $r=2(1+\cos\theta)$ (b) $r=2+3\cos\theta$

   (c) $r=3+2\cos\theta$

4   Sketch the spirals $r=2\theta$ and $r=\tfrac{1}{2}e^\theta$ for $0<\theta<2\pi$.

5   By writing $x=r\cos\theta$, $y=r\sin\theta$, express the following curves in polar form:

   (a) $x^2-y^2=a^2$ (b) $xy=c^2$ (c) $x^2+y^2+4x=0$

6   Express the following polar equations in cartesian form:

   (a) $r^2\sin 2\theta=1$ (b) $r^2\cos 2\theta=1$ (c) $r^2=a^2\sin 2\theta$

7   Find the area of the regions bounded by the following:

   (a)   the circle $r=a$

   (b)   the circle $r=a\cos\theta$

   (c) $r^2=a^2\sin 2\theta,\ -\pi/4<\theta<\pi/4$

   (d) $r^2=a^2\sin 2\theta,\ 0<\theta<\pi/2$

   (e) $r=a\theta,\ \theta=2\pi$

TRIGONOMETRY I

CONTENTS

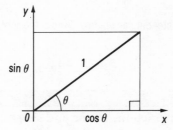

Fig 16.1

Cosine θ is defined as the projection on to the x-axis of a unit vector at an angle θ to Ox; **sine** θ as the projection on to the y-axis. From these definitions we see

$$\cos(-\theta)=\cos\theta=\cos(360°-\theta)=\cos(360°+\theta)$$

and $\quad \sin\theta=\sin(180°-\theta)=\sin(360°+\theta)=\sin(540°-\theta)$

Tan θ is defined as $\sin\theta/\cos\theta$

Some useful **exact** values:

$$\sin 0°=\cos 90°=0$$
$$\sin 30°=\cos 60°=\tfrac{1}{2}$$
$$\sin 60°=\cos 30°=\tfrac{1}{2}\sqrt{3}$$
$$\sin 90°=\cos 0°\ =1$$

Sine formula

$$\frac{a}{\sin A}=\frac{b}{\sin B}=\frac{c}{\sin C}=2R,\ R \text{ being the radius of the circumcircle of triangle } ABC.$$

$$a^2=b^2+c^2-2bc\cos A$$

Cosine formula

$$a^2=b^2+c^2-2bc\cos A$$

i.e. $\quad \cos A=\dfrac{b^2+c^2-a^2}{2bc}$

General solution of equations

If $\sin\theta=\sin\alpha$, $\theta=360n°+\alpha$ or $(360n+180n)°-\alpha$
If $\cos\theta=\cos\alpha$, $\theta=360n°\pm\alpha$
If $\tan\theta=\tan\alpha$, $\theta=180n°+\alpha$

Radian measure

An arc of a circle equal in length to the radius of the circle subtends an angle of **1 radian** at the centre of the circle. Thence

$$\pi\,\text{rad}=180°;\ 1\,\text{rad}=\left(\frac{180}{\pi}\right)°\approx57.3°$$

The length of an arc subtending an angle of θ rad at the centre is $r\theta$; the area of the sector, angle θ, is $\frac{1}{2}r^2\theta$.

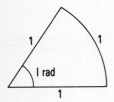

Fig 16.2

DEFINITION OF THE TRIGONOMETRIC FUNCTIONS

The trigonometric functions cosine and sine are best defined in terms of projections on to the coordinate axes. Thus if OP is a straight line unit length, inclined at an angle θ to Ox, cosine θ is defined as the projection of OP on to Ox, sine θ as the projection on to Oy. We can see from this definition that for any angle θ,

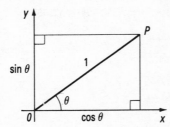

Fig 16.3

$$\cos(-\theta)=\cos\theta=(360°-\theta)=\cos(360°+\theta)\ldots$$

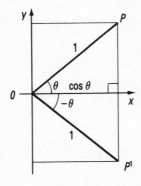

Fig 16.4(a)

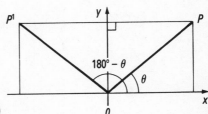

Fig 16.4(b)

and

$$\sin\theta=\sin(180°-\theta)=\sin(360°+\theta)=\sin(540°-\theta)\ldots$$

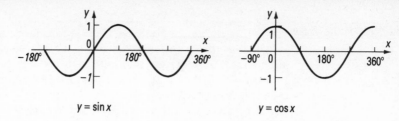

Fig 16.5 $y = \sin x$ $y = \cos x$

The graphs of sine and cosine are given in Fig 16.5.
Having defined cosine and sine, we now define

$$\tan\theta = \frac{\sin\theta}{\cos\theta}, \ \cot\theta = \frac{\cos\theta}{\sin\theta}$$

$$\sec\theta = \frac{1}{\cos\theta}, \ \operatorname{cosec}\theta = \frac{1}{\sin\theta}$$

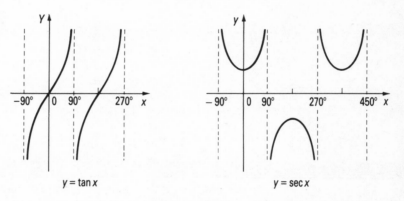

Fig 16.6 $y = \tan x$ $y = \sec x$

The graphs of tan and sec are given in Fig 16.6; the graphs of cot and
cosec are similar and can be deduced from these.

GENERAL SOLUTION OF THE EQUATION $\sin\theta = \sin\alpha$

Looking at the graph of $\sin\theta$ in Fig 16.5, we see that for any value k
such that $-1 \leqslant k \leqslant 1$, we have at least one value α such that $\sin\alpha = k$.
But we also see (Fig 16.7) that $\sin(180° - \alpha) = k = \sin\alpha$, so that if α is any
one solution (invariably taken in the first or fourth quadrant), of the
equation $\sin\alpha = k$, then further solutions are $180° - \alpha$, $360° + \alpha$, $540° - \alpha$
... and the general solution is

$$360n° + \alpha \text{ or } (2n+1)180° - \alpha$$

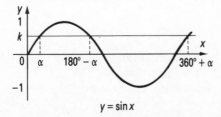

Fig 16.7 $y = \sin x$

GENERAL SOLUTION OF cos θ=cos α

Looking at the graph of cos θ in Fig 16.8, we see that for any value k such that $-1 \leqslant k \leqslant 1$, we have at least one value α such that cos $\alpha=k$. But we also see (Fig 16.8) that cos $(-\alpha)=k=$cos α, so that if α is any one solution (invariably taken in the first or second quadrant) to the equation cos $\alpha=k$, then further solutions are $360°-\alpha$, $360°+\alpha$, $720°-\alpha$, $720°+\alpha$ ... and the general solution is $360n°\pm\alpha$.

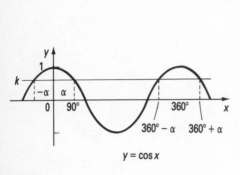

Fig 16.8

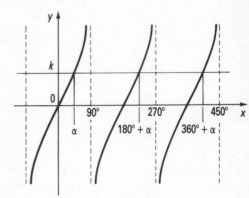

Fig 16.9 $y = \tan x$

GENERAL SOLUTION OF tan θ=tan α

From the graph of tan θ in Fig 16.9, we see that for any value k we can find first one solution α such that tan $\alpha=k$, then further values $180°+\alpha$, $360°+\alpha$, so that the general solution of the equation tan $\theta=$tan α, where tan $\alpha=k$ is $\theta=180n°+\alpha$, for any integer value of n.

Example 16.1 Find general solutions, in degrees, of (a)sin $\theta=-0.5$ (b) cos $\theta=0.5$ (c) tan $\theta=-1$.

(a)   From calculator or tables we obtain one solution of sin $\theta=-0.5$ as $\theta=-30°$. From the graph, we see that other solutions are 210°, 330°, 570°, ... and $-150°$, $-390°$, ... The general solution is
$$\theta=(360n-30)° \text{ or } [(2n+1)180+30]°$$

(b)   We first obtain one solution of cos $\theta=0.5$ as $\theta=60°$. From the graph, we see that other solutions are 300°, 420°, 660°, ... $-60°$, $-420°$ ... The general solution is
$$\theta=(360n\pm60)°$$

(c)   One solution of tan $\theta=-1$ is $\theta=-45°$. From the graph we see that other solutions are 135°, 315°, 495°, 675°, ... $-225°$, $-405°$, ... and the general solution is
$$\theta=(180n-45)°$$

RADIAN MEASURE

Although we are probably most familiar with the degree as the unit in which to measure angles, we shall certainly have come across radians, if only as a button on the face of a calculator! To define a radian, we say that if an arc of a circle, of unit radius, has length one unit, then the angle subtended by that arc at the centre of the circle is one radian, written 1 rad. Since the circumference of this circle is 2π, the angle subtended by the complete circumference is 2π rad, so that

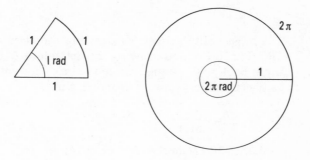

Fig 16.10

$$360° = 2\pi \text{ rad}$$

Using this relation, we see, for example,

$$60° = \pi/3 \text{ rad}, 150° = 5\pi/6 \text{ rad}, 270° = 3\pi/2 \text{ rad}$$

and, using calculators,

$$50° \approx 0.873 \text{ rad}, 150° \approx 2.62 \text{ rad and } 250° \approx 4.36 \text{ rad}$$

Similarly,

$$\pi/2 \text{ rad} = 90°, 2\pi/3 \text{ rad} = 120°,$$

and by calculators,

$$1 \text{ rad} \approx 57.3, 1.4 \text{ rad} = 1.4 \times 180/\pi \approx 80°$$

SINE FORMULA

The sine formula states that, in any triangle ABC,

$$\frac{a}{\sin A} = \frac{b}{\sin B} = \frac{c}{\sin C} = 2R$$

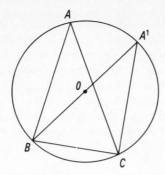

Fig 16.11

It is most easily proved by drawing the circumcircle of the triangle ABC (Fig 16.11), and drawing the diameter BA' through B. Then since angle $BCA'=90°$ (angle in a semi-circle) and angle $BA'C$=angle BAC (angles in same segment), $BC=BA'\sin A$,

i.e. $$2R=\frac{a}{\sin A}$$

Similarly, $2R=b/\sin B$ and $2R=c/\sin C$, so

$$\frac{a}{\sin A}=\frac{b}{\sin B}=\frac{c}{\sin C}=2R$$

COSINE FORMULA

The cosine formula states that in any triangle ABC,

$$a^2=b^2+c^2-2bc\cos A$$

If we wish to find an angle knowing the three sides of the triangle, then we can rearrange the formula

$$\cos A=\frac{b^2+c^2-a^2}{2bc}$$

Similar formulae exist with b, c, $\cos B$ or $\cos C$ as the subject of the formula.

To prove the cosine formula, draw a line through B perpendicular to AC, meeting AC, produced if necessary, at D (see Fig 16.12(b)). Then $AD=c\cos A$, and $DC=b-c\cos A$ (Fig 16.12(a)).

Since $$BD^2=BC^2-DC^2$$
and $$BD^2=BA^2-AD^2$$
$$BC^2-DC^2=BA^2-AD^2$$
i.e. $$a^2-(b-c\cos A)^2=c^2-(c\cos A)^2$$
whence $a^2-b^2+2bc\cos A-c^2\cos^2 A=c^2-c^2\cos^2 A$

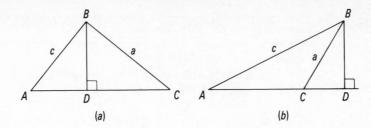

Fig 16.12 (a) (b)

i.e. $a^2=b^2+c^2-2bc\cos A$

The proof needs to be adapted slightly if angle A or C is obtuse. We are certainly familiar with both these formulae, but is may be useful to be reminded of special cases where we may have some ambiguity. Since, e.g. $\sin 30°=\sin 150°=0.5$, we may have two possible values for any angle; often only one of these is correct.

Example 16.2 In the triangle ABC, the length of BC is 4cm, the length of AC is 7cm and $C=40°$. Find the remaining sides and angles.

Using the cosine formula first,

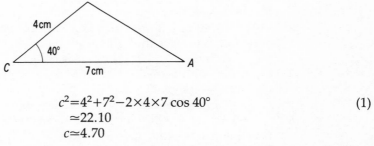

Fig 16.13

$$c^2=4^2+7^2-2\times4\times7\cos 40° \qquad (1)$$
$$\approx22.10$$
$$c\approx4.70$$

Now calculate the angle opposite the smaller of the two given sides, as this cannot be obtuse.

$$\sin A=\frac{4\sin 40°}{4.70} \qquad (2)$$

$$A=33.2°$$

Since the angle-sum of the triangle is 180°, $B=106.8°$. The length of AB is 4.70cm, angle $A=33.2°$ and angle $B=106.8°$.
N.B. Take particular care when evaluating expressions like (1) and (2) with a calculator; always make a rough estimate of the expected answer.

QUESTIONS

1 Using graph paper, draw the graph of $y=\cos\theta$, for values of θ between $-360°$ and $+360°$.

 Using this graph, draw sketch-graphs of the following:

   (a) $y=2\cos\theta$ (b) $y=2+\cos\theta$

   (c) $y=\cos(\theta+30°)$

2 As in Q.1, draw the graph of $y=\sin\theta$, for values of between $-360°$ and $+360°$.

 Using this graph, draw sketch graphs of the following:

   (a) $y=2\sin\theta$ (b) $y=2+\sin\theta$

   (c) $y=\sin(\theta+30°)$

3 First, from calculator or tables, find one solution in degrees to each of the following, then deduce from graphs the general solutions:

   (a) $\sin\theta=0.8$ (b) $\cos\theta=-0.9$ (c) $\tan\theta=2$

   (d) $\tan\theta=-1.5$ (e) $\cos\theta=0.7$ (f) $\sin\theta=-0.4$

4 Express the following angles in radian measure giving your answers as multiples of π:

   (a) $45°$ (b) $210°$ (c) $-90°$

   (d) $240°$

5 Express the following angles in radian measure, correct to 3 s.f.:

   (a) $40°$ (b) $200°$ (c) $-10°$

   (d) $-100°$

6 First find, in radian measure as a multiple of π, one solution to each equation, then give the general solution:

   (a) $\sin\theta=1$ (b) $\cos\theta=0.5$ (c) $\tan\theta=1$

7 Find, in radian measure correct to 3 d.p., one solution of each of the following equations:

   (a) $\sin\theta=0.3$ (b) $\cos\theta=-0.6$ (c) $\tan\theta=-2$

8 In the triangle ABC, $a=3$cm, $b=5$cm and angle $C=45°$. Find the remaining sides and angles.

9 In the triangle ABC, $a=6$cm, $b=7$cm and $c=8$cm. Find the angles of the triangle.

10 A glider pilot at a point P, 1000m above sea-level, observes two beacons at points Q and R. He estimates that his horizontal distance from Q is 3000m, its angle of elevation from P is $10°$ and its bearing from P is 020. He estimates his horizontal distance from R is 1000m, its angle of depression is $15°$ and its bearing from P is 080. Find the heights of Q and R above sea-level, and the angle of elevation of Q from R.

TRIGONOMETRY II

CONTENTS

SUMS AND PRODUCTS FORMULAE

$$\sin(A+B)= \sin A \cos B+\cos A \sin B$$
$$\sin(A-B)=\sin A \cos B-\cos A \sin B$$
$$\cos(A+B)=\cos A \cos B-\sin A \sin B$$
$$\cos(A-B)=\cos A \cos B+\sin A \sin B$$

$$\tan(A+B)= \frac{\tan A+\tan B}{1-\tan A \tan B}$$

$$\sin 2A=2 \sin A \cos A$$
$$\cos 2A=\cos^2 A-\sin^2 A$$
$$=2 \cos^2 A-1$$
$$=1-2 \sin^2 A$$

$$\tan 2A= \frac{2 \tan A}{1-\tan^2 A}$$

$$\sin x+\sin y=2 \sin \tfrac{1}{2}(x+y) \cos \tfrac{1}{2}(x-y)$$
$$\sin x-\sin y=2 \cos \tfrac{1}{2}(x+y) \sin \tfrac{1}{2}(x-y)$$
$$\cos x+\cos y=2 \cos \tfrac{1}{2}(x+y)\cos \tfrac{1}{2}(x-y)$$
$$\cos x-\cos y=-2 \sin \tfrac{1}{2}(x+y)\sin \tfrac{1}{2}(x-y)$$

HALF ANGLE FORMULAE

If $t=\tan \tfrac{1}{2}A$,

$$\sin A=\frac{2t}{1+t^2}$$

$$\cos A=\frac{1-t^2}{1+t^2}$$

$$\tan A=\frac{2t}{1-t^2}$$

AUXILIARY ANGLE

$$a \sin \theta+b \cos \theta=R \sin (\theta+\alpha)$$

where $R=\sqrt{a^2+b^2}$, and $\cos \alpha$: $\sin \alpha$:$1=a$:b:R.

$$-\sqrt{(a^2+b^2)}\leq a \sin \theta+b \cos \theta \leq \sqrt{(a^2+b^2)}$$

for all values of θ.

SUMS AND PRODUCTS FORMULAE

Certain formulae must be learnt:

$$\sin (A+B)=\sin A \cos B+\cos A \sin B \tag{1}$$
$$\sin (A-B)=\sin A \cos B-\cos A \sin B \tag{2}$$
$$\cos (A+B)=\cos A \cos B-\sin A \sin B \tag{3}$$
$$\cos (A-B)=\cos A \cos B+\sin A \sin B \tag{4}$$

The proofs are rarely required (some examination syllabuses specifically state that the proofs are *not* required), but proofs can be found in most level A textbooks.*

*See e.g. *Additional Pure Mathematics* by L. Harwood Clarke and F. G. J. Norton

Adding (1) and (2), we have

$$\sin (A+B)+\sin(A-B)=2 \sin A \cos B \tag{5}$$

and writing $A+B=x$, $A-B=y$

$$\sin x+\sin y=2 \sin \tfrac{1}{2}(x+y) \cos \tfrac{1}{2}(x-y)$$

which is most easily remembered as

$$\sin+\sin=2 \sin \text{ (half sum) } \cos \text{ (half difference)}$$

where difference means 'first minus second'.
Similarly we obtain

$$\sin x-\sin y=2 \cos \tfrac{1}{2}(x+y) \sin \tfrac{1}{2}(x-y) \tag{6}$$
$$\text{'sin}-\text{sin}=2 \cos \text{ (half sum)sin(half difference)'}$$
$$\cos x+\cos y=2 \cos \tfrac{1}{2}(x+y) \cos \tfrac{1}{2}(x-y) \tag{7}$$
$$\text{'cos}+\text{cos}=2 \cos\text{(half sum)cos(half difference)'}$$

and

$$\cos x-\cos y=-2 \sin \tfrac{1}{2}(x+y)\sin\tfrac{1}{2}(x-y) \tag{8}$$
$$\text{'cos}-\text{cos}=-2 \sin \text{ (half sum)sin(half difference)'}$$

N.B. *Remember the minus sign in (8).*

DOUBLE ANGLE FORMULAE

From (1) and (3) we can deduce the double angle formulae,

$$\sin 2A=2 \sin A \cos A \tag{9}$$
$$\cos 2A=\cos^2 A-\sin^2 A \tag{10}$$
$$=2 \cos^2 A-1, \text{ using } \sin^2 A=1-\cos^2 A$$
$$=1-2 \sin^2 A$$

From (1) and (3) we also have

$$\tan (A+B)=\frac{\tan A+\tan B}{1-\tan A \tan B} \tag{11}$$

whence

$$\tan 2A = \frac{2 \tan A}{1 - \tan^2 A} \tag{12}$$

These formulae are often useful when solving equations or establishing identities.

Example 17.1 Solve the equation, for values of x between $0°$ and $360°$,

$$\sin x + \sin 2x = \cos \tfrac{1}{2}x$$

From (6), we see $\sin x + \sin 2x = 2 \sin \tfrac{3}{2}x \cos \tfrac{1}{2}x$ so that the equation becomes

$$2 \sin \tfrac{3}{2}x \cos \tfrac{1}{2}x = \cos \tfrac{1}{2}x$$

i.e. either $\cos \tfrac{1}{2}x = 0$ or $2 \sin \tfrac{3}{2}x = 1$.

If $\cos \tfrac{1}{2}x = 0$, $\tfrac{1}{2}x = 90°$ or $270°$, so $x = 180°$ is the only solution in the interval $0 \leqslant x \leqslant 360$.
If $2 \sin \tfrac{3}{2}x = 1$, $\sin \tfrac{3}{2}x = \tfrac{1}{2}$,

$$\tfrac{3}{2}x = 30° \text{ or } 150° \text{ or } 390° \text{ or } 510°$$
$$x = 20° \text{ or } 100° \text{ or } 260° \text{ or } 340°$$

so the complete list of solutions is

$$x = 20° \text{ or } 100° \text{ or } 180° \text{ or } 260° \text{ or } 340°$$

Check: when $x = 20°$

$$\sin 20° + \sin 40° \approx 0.98$$
$$\cos 10° \approx 0.98$$

Notice that to ensure we had all solutions in the range $0°$ to $360°$ we had to consider values of $\tfrac{3}{2}x$ greater than $360°$, indeed all values up to $\tfrac{3}{2}x \times 360°$, since if $\tfrac{3}{2}x < 540°$, $x < 360°$.

IDENTITIES

When establishing identities it is most important that we should not begin by assuming what we wish to prove, for

'If $7 = 5$,
$\quad 5 = 7$,
$\quad 12 = 12$, which is true'

yet it is plain that the original premise is false!

Almost invariably, consider the left hand side of the identity and express it in the same form as the right hand side.

Example 17.2 If $\cos \theta \neq -\frac{1}{2}$ or 0, show that

$$\frac{\sin \theta + \sin 2\theta}{1 + \cos \theta + \cos 2\theta} = \tan \theta$$

First notice that the right hand side is $\dfrac{\sin \theta}{\cos \theta}$, so that we hope to show that $\sin \theta$ is a factor of the numerator of the left hand side, and $\cos \theta$ is a factor of the denominator. This suggests that we shall first try using

$$\sin 2\theta = 2 \sin \theta \cos \theta, \qquad\qquad (9)$$
$$1 + \cos 2\theta = 2 \cos^2 \theta \qquad\qquad \text{from (10)}$$

so that the left hand side (L.H.S) is

$$\frac{\sin \theta + 2 \sin \theta \cos \theta}{\cos \theta + 2 \cos^2 \theta}$$

$$= \frac{\sin \theta (1 + 2 \cos \theta)}{\cos \theta (1 + 2 \cos \theta)}$$

$$= \tan \theta, \text{ since } 1 + 2 \cos \theta \neq 0$$

We may have been tempted to try

$$\sin \theta + \sin 2\theta = 2 \sin \tfrac{3}{2}\theta \cos \tfrac{1}{2}\theta$$

or even

$$1 + \cos \theta + \cos 2\theta = 1 + 2 \cos \tfrac{3}{2}\theta \cos \tfrac{1}{2}\theta$$

but the ratio $\tfrac{1}{2}\theta$ is not like any term in the R.H.S., so that this method does not look worth pursuing. **All identities set at A-level nowadays are straightforward and can be established quickly, so do not get involved in heavy algebra!**

These formulae are often useful when integrating expressions (p242).

HALF-ANGLE FORMULAE

From equation (9), we have

$$\sin A = 2 \sin \tfrac{1}{2}A \cos \tfrac{1}{2}A$$

$$= \frac{2 \sin \tfrac{1}{2}A \cos \tfrac{1}{2}A}{\cos^2 \tfrac{1}{2}A + \sin^2 \tfrac{1}{2}A}$$

$$= \frac{2 \tan \tfrac{1}{2}A}{1 + \tan^2 \tfrac{1}{2}A}$$

which we can write as $\quad \sin A = \dfrac{2}{1 + t^2}$, where $t = \tan \tfrac{1}{2}A \quad (13)$

Similarly we can show

$$\cos A = \frac{1-t^2}{1+t^2} \tag{14}$$

and

$$\tan A = \frac{2t}{1-t^2} \tag{15}$$

Equations (14) and (15) are especially useful for solving some equations, and for finding certain integrals.

Example 17.3 Solve correct to 1 d.p. the equation $3 \sin x + 4 \cos x = 2$ for $0° < x < 360°$.

Method 1: Using (13) and (14) we have

$$3\frac{2t}{1+t^2} + 4\frac{1-t^2}{1+t^2} = 2$$

i.e. $\quad\quad\quad\quad 6t + 4 - 4t^2 = 2 + 2t^2$

i.e. $\quad\quad\quad\quad 3t^2 - 3t - 1 = 0$

$\tan \frac{1}{2}x \equiv t = 1.26$ or -0.26, using the formula for the solution of quadratic equations,

$$\tfrac{1}{2}x = 51.65° \text{ or } 165.22°$$
$$x = 103.3° \text{ or } 330.4°$$

Notice that to find solutions between 0° and 360° we only had to consider values of $\frac{1}{2}x$ between 0° and 180°, but in order to obtain solutions correct to 1 d.p., we had to find $\frac{1}{2}x$ to a greater degree of accuracy.

Method 2: Comparing $3 \sin x + 4 \cos x$ with $\sin x \cos \alpha + \cos x \sin \alpha$ we see that if we write

$$3 \sin x + 4 \cos x = 5(\sin x \cos \alpha + \cos x \sin \alpha)$$
$$= 5 \sin (x + \alpha)$$

our given equation reduces to

$$5 \sin (x + \alpha) = 2$$
i.e. $\quad \sin (x + \alpha) = 0.4$

α being such that $\cos \alpha = 0.6$, $\sin \alpha = 0.8$, i.e. $\alpha \simeq 53.13°$.

Thus $\sin (x + 53.13°) = 0.4$
$$x + 53.13° = 23.58° \text{ or } 156.42° \text{ or } 383.58° \text{ or } \dots$$

the values of x between 0° and 360° being 103.3° or 330.4°. Notice that to obtain values of x in the required range we had to consider values of $x + \alpha$ in the range 53.13° to 413.13°, and also that, as before, we worked to 2 d.p. to obtain a final answer correct to 1 d.p. Much of the arithmetic done in detail above should be carried out on a calculator.

AUXILIARY ANGLE

The angle α we introduced above is called the auxiliary angle. The form (16) is useful not only for solving equations of this type, but it also enables us to see that

$$-5 \leqslant 3 \sin x + 4 \cos x \leqslant 5$$
since $\quad -5 \leqslant 5 \sin(x+\alpha) \leqslant 5$

the maximum value of 5 being attained when $x=90°-\alpha$, the minimum value when $x=270°-\alpha$.

QUESTIONS

1 Express the following as products of two trig ratios:
   (a) $\sin 5x + \sin x$ (b) $\cos 5x + \cos 3x$
   (c) $\sin 5x - \sin 2x$ (d) $\cos 5x - \cos 7x$

2 Express the following as the sum of two trig ratios:
   (a) $2 \sin 5x \cos 3x$ (b) $2 \cos 4x \cos x$
   (c) $\sin 4x \cos 5x$ (d) $\sin x \sin 2x$

3 Writing $\cos 4x = \cos(2x+2x)$. Express $\cos 4x$ as a polynomial in $\cos x$.

4 Solve the following equations, for values of x such that $0° \leqslant x \leqslant 360°$.
   (a) $2 \sin x + \sin 2x = 0$
   (b) $\sin x + \sin 3x + \sin 5x = 0$
   (c) $2 \cos 2x + 2 \sin^2 x - 1 = 0$
   (d) $\tan 2x + 5 \tan x = 0$

5 Establish the following identities:
   (a) $\cos x + \cos(x+120°) + \cos(x+240°) = 0$
   (b) $\sin 2x - \sin 4x + \sin 6x = \sin 4x(2 \cos 2x - 1)$
   (c) $\sin^2 x - \sin^2 y = \sin(x+y) \sin(x-y)$

   (d) $\dfrac{1}{1-\cos x} + \dfrac{1}{1+\cos x} = 2 \operatorname{cosec}^2 x$

6 Given that $\sin x = 0.8$, $\cos x = -0.6$, find the exact value of the following:
   (a) $\sin 2x$
   (b) $\cos 4x$
   (c) $\sin(x+45°)$
   (d) $\sin(x+120°) + \sin(x-120°)$

   Solve the following equations, correct to the nearest degree, for values of x such that $0° < x < 360°$:
7 $3 \sin x + 4 \cos x = 1$
8 $\sin x + \cos x = 0.5$
9 $3 \cos x - 2 \sin x = 1$
10 $3 \sin x - 5 \cos x = 5$
11 $3 \sin 2x + 4 \cos 2x = 1$

COMPLEX NUMBERS

CONTENTS

COMPLEX CONJUGATE

If $z=a+ib$, the complex conjugate of z, written z^* (or $\bar{z}$) is $\bar{z}=a-ib$.

MODULUS AND ARGUMENT

The modulus of $|z|$, written $|z|$, is $\sqrt{(a^2+b^2)}$.
The argument of z, written arg z, is arc tan (b/a), **only if x is positive**. Draw a diagram for all other cases.

PRODUCTS AND QUOTIENTS

$$R(\cos\theta+i\sin\theta)(r\cos\varphi+i\sin\varphi)$$
$$=Rr[\cos(\theta+\varphi)+i\sin(\theta+\varphi)]$$

$$\frac{R(\cos\theta+i\sin\theta)}{r(\cos\varphi+i\sin\varphi)}=\frac{R}{r}[\cos(\theta-\varphi)+i\sin(\theta-\varphi)]$$

$$[R(\cos\theta+i\sin\theta)]^2=R^2(\cos2\theta+i\sin2\theta)$$

LOCI

$|z|=1$ represents a circle, centre $(0,0)$, radius 1.
$|z-a-ib|=r$ represents a circle, centre (a,b), radius r.
arg $z=\alpha$ is a straight line through $(0,0)$, inclined at an angle α to the real axis.
$|z-z_1|=|z-z_2|$ is the perpendicular bisector of the straight line joining the points representing the complex numbers z_1, z_2.

Complex numbers can be regarded merely as an extension of the real number system, necessary if we are to consider the solution of certain equations. Just as a child may say 'there are no solutions to the equation $3x=4$' until he has been introduced to 'fractions', so we may say 'there are no solutions to the equation $x^2=-4$' until we are introduced to the number i such that $i\times i=-1$. What is remarkable is that this extension of the number-field enables us to solve all polynomials, even those with complex coefficients, and also equations like $3^x=-2$ and $\sin x=-3$.

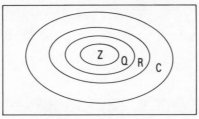

Z = {integers}
Q = {rationals}
R = {reals}
C = {complex numbers}

Fig 18.1

COMPLEX CONJUGATE

Associated with each complex number $z=a+ib$ we define its complex conjugate z^* (sometimes written $\bar{z}$), where $z^*=a-ib$. This has the property that

$$z+z^*=(a+ib)+(a-ib)=2a, \text{ which is real}$$
and $$zz^*=(a+ib)(a-ib)=a^2+b^2, \text{ which is also real}$$

OPERATIONS WITH COMPLEX NUMBERS

Addition and subtraction are defined by

$$(a+ib)+(c+id)=(a+c)+i(b+d)$$
and $$(a+ib)-(c+id)=(a-c)+i(b-d)$$

i.e. add (or subtract) the real and imaginary parts separately:

multiplication is defined by

$$(a+ib)(c+id)=ac-bd+i(bc+ad)$$

N.B. Take special care that $ib\times id=-bd$; although we may be confident that we will never make a mistake, it is terribly easy to write $ib\times id=+bd$.

To divide two complex numbers, we almost invariably multiply numerator and denominator by the complex conjugate of the denominator, as in this example.

Example 18.1 If $z_1=2+i$ and $z_2=4+3i$, find, in the form $a+ib$

$(a)\dfrac{z_1}{z_2}(b)2+1/z_1$

To express $\dfrac{z_1}{z_2}$ in this form,

$$\frac{z_1}{z_2} = \frac{2+i}{4+3i} = \frac{(2+i)(4-3i)}{(4+3i)(4-3i)}$$

$$= \frac{8-6i+4i+i\times(-3i)}{16+9}$$

$$= \tfrac{11}{25} - \tfrac{3}{25}i$$

Similarly, $\quad 2 + \dfrac{1}{z_1} = 2 + \dfrac{2-i}{(2+i)(2-i)}$

$$= 2 + \tfrac{2}{5} - \tfrac{1}{5}i$$

$$= \tfrac{12}{5} - \tfrac{1}{5}i$$

adding the two real numbers.

EQUALITY OF COMPLEX NUMBERS

Two complex numbers are equal only if the real parts of each are equal and the imaginary parts of each are equal, thus if

$$x+iy=2+3i, \ x=2 \text{ and } y=3$$

REPRESENTATION IN AN ARGAND DIAGRAM

In the same way that real numbers can be represented along

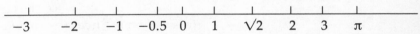

a number-line, we can represent complex numbers using two dimensions so that a point (x,y) represents the complex number $x+iy$. The distance of the point P, representing the complex number z, from the origin is called the modulus of z, written $|z|$; the angle made by OP with the positive x-axis is called the argument (the term amplitude was used until fairly recently).

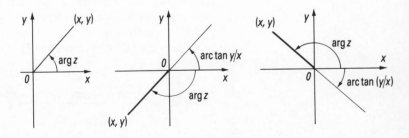

Fig 18.2

Thus if $z=x+iy$, $|z|=\sqrt{(x^2+y^2)}$
arg $(z)=$ arc tan (y/x) if x is positive
$=$ arc tan $(y/x)-\pi$ if both x and y are negative,
$=$ arc tan $(y/x)+\pi$ if x is negative but y is positive.

Always draw a diagram when finding the argument of a complex number.

Example 18.2 Find the modulus and argument of (a) $1+i$ (b) $-1-i$ (c) $-1+i$

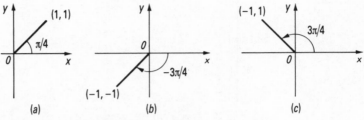

Fig 18.3

(a)　　　　　　　(b)　　　　　　　(c)

In each case, the modulus is $\sqrt{2}$.
From Fig 18.3(a) we see arg $(1+i)=$ arc tan $(1/1)=\pi/4$.
From Fig 18.3(b) we see arg $(-1-i)=\pi/4-\pi=-3\pi/4$.
From Fig 18.3(c) we see arg $(-1+i)=-\pi/4+\pi=3\pi/4$.

PRODUCT AND QUOTIENT OF TWO COMPLEX NUMBERS

The modulus-argument form of a complex number enables us to divide by a complex number more easily than using the $a+ib$ form. For if $z=r(\cos\theta+i\sin\theta)$, r being the modulus and θ the argument.

$$\frac{1}{z}=\frac{1}{r(\cos\theta+i\sin\theta)}=\frac{\cos\theta-i\cos\theta}{r(\cos\theta+i\sin\theta)(\cos\theta-i\sin\theta)}$$

$$=\frac{(\cos\theta-i\sin\theta)}{r}, \text{ since } \cos^2\theta+\sin^2\theta=1$$

$$=\frac{\cos(-\theta)+i\sin(-\theta)}{r}$$

Also, if $z_1=r_1(\cos\theta+i\sin\theta)$, $z_2=r_2(\cos\varphi+i\sin\varphi)$,

$$z_1z_2=r_1r_2(\cos\theta+i\sin\theta)(\cos\varphi+i\sin\varphi)$$
$$=r_1r_2(\cos\theta\cos\varphi-\sin\theta\sin\varphi+i\sin\theta\cos\varphi$$
$$+i\cos\theta\sin\varphi)$$
$$=r_1r_2[\cos(\theta+\varphi)+i\sin(\theta+\varphi)]$$

and

$$\frac{z_1}{z_2}=z_1\times\frac{1}{z_2}=\frac{r_1(\cos\theta+i\sin\theta)[\cos(-\varphi)+\sin(-\varphi)]}{r_2}$$

$$=\frac{r_2}{r_2}[\cos(\theta-\varphi)+i\sin(\theta-\varphi)]$$

Thus to multiply two complex numbers in modulus-argument form, we multiply the moduli and add the arguments; to divide, we divide the moduli and subtract the arguments.

POWERS OF COMPLEX NUMBERS

This result when multiplying two complex numbers gives us an easy way of finding the powers of a complex number, for

$$z^2 \equiv [r(\cos\theta + i\sin\theta)]^2 = r^2(\cos\theta + i\sin\theta)(\cos\theta + i\sin\theta)$$
$$= r^2(\cos\theta + i\sin 2\theta)$$

and $\quad z^3 \equiv z^2 z = r^2(\cos 2\theta + i\sin 2\theta)r(\cos\theta + i\sin\theta)$

$$= r^3(\cos 3\theta + i\sin 3\theta)$$

We can prove by induction, that for all positive integer values of n,

$$z^n \equiv [r(\cos\theta + i\sin\theta)]^n = r^n(\cos n\theta + i\sin n\theta)$$

This theorem, known as de Moivre's theorem, is true for all values of n, whether integer or not, but the general form is outside the syllabus of single-subject A-level examinations.

LOCI

All points whose distance from the fixed point O is constant, say 2 units, lie on a circle, centre O, radius 2. Thus $|z| = 2$ describes a circle, centre O, radius 2.

All points P such that OP is inclined to the positive x-axis at a constant angle, say $\pi/6$, lie on a straight line through O inclined at $\pi/6$ to the positive x-axis. (Strictly, they lie on only half the straight line, since the argument of numbers represented by points on the other half is $-5\pi/6$.)

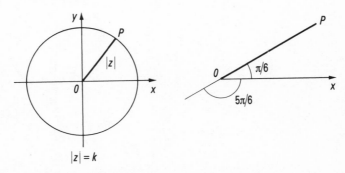

Fig 18.4 $|z| = k$

If the point P represents the complex number z, Fig 18.5 shows the relation of P to the point Q representing the number $z-2$, and to the point R representing the number $z-1+2i$.

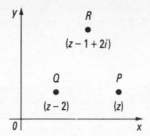

Fig 18.5

Thus if P is such that $|z-2|=1$, the point representing the complex number $Z\equiv z-2$ describes as circle centre O, radius 1, so that the point P representing z, i.e. $Z+2$, describes a circle centre $(2,0)$, radius 1. Similarly if $|z-1+2i|=3$, P describes a circle, centre $(1, -2)$ radius 3. We may be helped in interpreting these loci if we remember the relation of, say, $z-2$ to z (Fig 18.6), so that if the point representing $z-2$ describes a circle centre the origin, the point representing z will describe this circle after the point has been translated by $+2$; if $z-1+2i$ describes a circle centre O, the point z describes this circle after a translation described by $1-2i$.

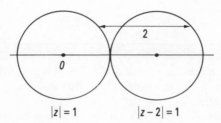

Fig 18.6

Example 18.3 Find the locus described by $|z-i|=2$

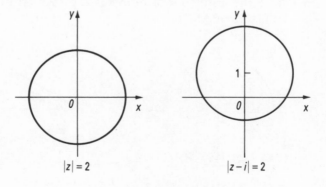

Fig 18.7

The point P' representing the number $Z\equiv z-i$ describes a circle, centre O, radius 2. The point P representing the number z, i.e. $(z-1)+i\equiv Z+i$, is P' **after a translation**$+i$, so P describes a circle, centre $(0,1)$, radius 2.

Example 18.4 Find the locus described by arg $(z-1)=\pi/4$

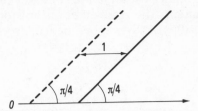

Fig 18.8

We know that arg $Z=\pi/4$ is half of a straight line through O, so if

$$Z=z-1$$
$$z=Z+1$$

and the locus of z will be the locus of Z **after a translation of +1**.

Example 18.5 Find the locus described by $|z-1-i|=|z+1|$

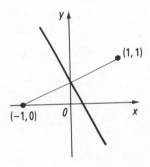

Fig 18.9

$|z-1-i|$ is the distance of the point P from the point representing the number $1+i$, i.e. from $(1,1)$; $|z+1|$ is the distance from the point representing the number -1, i.e. $(-1,0)$. The locus of points equidistant from points A and B is the perpendicular bisector of AB, so the locus described by

$$|z-1-i|=|z+1|$$

is the perpendicular bisector of the line joining $(1,1)$ to $(-1,0)$.

ALGEBRAIC METHOD

Many of these loci can be found algebraically, though this is usually much longer than using the geometric property stated in the equation.

In Example 18.5, since $|z|=\sqrt{x^2+y^2}$

$$|z-1-i|=|z+1|$$
$$\Rightarrow|x+iy-1-i|=|x+iy+1|$$
$$(x-1)^2+(y-1)^2=(x+1)^2+y^2$$
$$x^2-2x+1+y^2-2y+1=x^2+2x+1+y^2$$

i.e. $\qquad\qquad\qquad\qquad\qquad 4x+2y=1$

We recognize this as a straight line, but may not easily see its geometrical interpretation. The gradient of the line joining $(-1,0)$ to $(1,1)$ is $\frac{1}{2}$; its midpoint is $(0,\frac{1}{2})$, so the perpendicular bisector has equation

$$y-\tfrac{1}{2}=-2(x-0)$$

i.e. $\qquad\qquad\qquad\qquad\qquad 4x+2y=1$

QUESTIONS

1 If $z=2-i$, find in the form $a+ib$

    (a) $\qquad z+2i$ (b) $\qquad z^2$ (c) $\qquad \bar{z}$

    (d) $\qquad 1/z$ (e) $\qquad \dfrac{1}{z+1}$

2 If $z_1=3+2i$, $z_2=1-3i$, find in the form $a+ib$,

    (a) $\qquad z_1+z_2$ (b) $\qquad z_1 z_2$ (c) $\qquad z_1 \bar{z}_2$

    (d) $\qquad \dfrac{1}{z_1}+\dfrac{1}{z_2}$ (e) $\qquad \dfrac{1+z_1}{z_2-2i}$

3 If $z_1=3+2i$ and $z_2=1+3i$, find real numbers a, b, c, d such that

    (a) $\qquad az_1+bz_2=9-i$

    (b) $\qquad \dfrac{c}{z_1}+\dfrac{d}{z_2}=9-13i$

4 If $\dfrac{z}{z-2}=2+i$, find z in the form $a+ib$

5 If $z=3-2i$, find $\bar{z}$, $z\bar{z}$ and $z+\bar{z}$. Hence obtain the quadratic equation whose roots are z and $\bar{z}$.

6 Represent on an Argand diagram, and give the modulus and argument of

    (a) $\qquad 3$ (b) $\qquad 2i$ (c) $\qquad -1$
    (d) $\qquad -2i$ (e) $\qquad 3+4i$

7 Represent on an Argand diagram, and give the modulus and argument of

    (a) $\qquad \frac{1}{2}+i\dfrac{\sqrt{3}}{2}$ (b) $\qquad \sqrt{3}-i$ (c) $\qquad -2\sqrt{3}+2i$

    (d) $\qquad -\frac{1}{2}-i\dfrac{\sqrt{3}}{2}$ (e) $\qquad -\sqrt{3}-i$

8 If $\quad z_1-2(\cos \pi/3+i\sin \pi/3)$
        $z_2=5(\cos \pi/6+i\sin \pi/6)$
    and $z_3=3(\cos \pi/4-i\sin \pi/4)$
    find, in modulus-argument form

    (a) $\qquad z_1 z_2$ (b) $\qquad \dfrac{z_1}{z_2}$ (c) $\qquad z_1 z_2 z_3$

    (d) $\qquad z_1^2$ (e) $\qquad z_2^3$ (f) $\qquad \dfrac{z_1 z_2}{z_3^2}$

9 If $z = 3/2 + i\dfrac{\sqrt{3}}{2}$, write z in modulus argument form. Illustrate this form in an Argand diagram, and use it to find
   (a) z^2 (b) z^3 (c) $1/z$
   (d) the smallest positive value of n for which z^n is real.

10 If $z = 1 + \cos\theta + i\sin\theta$, write z in modulus-argument form. Illustrate this form in an Argand diagram, and use it to find
   (a) z^2 (b) $1/z$ (c) z^4

11 Sketch the locus $|Z| = 2$. Writing $Z = z - 2 - 3i$, i.e. $z = Z + 2 + 3i$, deduce the locus $|z - 2 - 3i| = 2$. Show this locus in the same sketch as the first locus.

12 Sketch the loci described by
   (a) $|z - 2| = 1$ (b) $|z - 2i| = 1$
   (c) $|z + 2i| = 1$ (d) $|z - 2 - 2i| = 1$

13 Sketch the loci described by
   (a) $\arg(z - 1) = \pi/3$ (b) $\arg(z + 1) = \pi/3$
   (c) $\arg(z - 1) = -2\pi/3$ (d) $\arg(z + i) = \pi/3$

14 Sketch the loci described by
   (a) $|z| = |z - 2|$ (b) $|z - i| = |z + 2|$

15 Sketch the loci $\arg z = \pi/4$ and $|z + 2 + i| = |z - 4 + i|$. Use geometry to find the coordinates of the point common to both loci.

16 If $z_1 = 10 - 2i$ and $z_2 = 2 - 3i$, show that $\arg(z_1/z_2) = \pi/4$.

17 If P represents the complex number $\sqrt{3} + i$, find geometrically the two possible complex numbers represented by Q, the third vertex of the equilateral triangle OPQ.

18 Find, each in the form $a + ib$, the sum and the product of the roots of the quadratic equation.
   $$(1 - i)z^2 - 2iz + 3 - i = 0$$

19 If z is such that $z^3 = 1$ but $z \neq 1$, show that $z^2 + z + 1 = 0$.

20 A square is inscribed in the circle $|z| = 2$, so that the four vertices lie on the circumference of the circle; one of its vertices represents the complex number $\sqrt{2} + i\sqrt{2}$. Find the numbers represented by the other vertices.

21 A point P describes the circle $|z| = 2$ in an anti-clockwise sense. Describe the loci
   (a) $|z| = 2$ (b) $|1/z| = 2$

22 Sketch in a diagram the region described by
   $$|z - 1| \leqslant 1$$

   and $\qquad \dfrac{-\pi}{4} \leqslant \arg z \leqslant \dfrac{\pi}{4}$

23 Find the roots of the equation $z^2 + 2z + 5 = 0$. Show that for any quadratic equation with real coefficients that has complex roots, the roots are complex conjugates.

24 Expand $(\cos\theta + i\sin\theta)^5$ by the binomial theorem. From this expansion, show that $\cos 5\theta = 16c^2 - 20c^3 + 5c$, where $c = \cos\theta$. Obtain a similar expression for $\cos 6\theta$.

25   If $|z_1 - z_2| = |z_1 + z_2|$, show geometrically that the arguments of z_1 and z_2 differ by $\pi/2$.

MATRICES

CONTENTS

TRANSPOSE OF A MATRIX

If $\mathbf{A} = \begin{pmatrix} a\,b\,c \\ x\,y\,z \end{pmatrix}$, the transpose, $\mathbf{A}^T = \begin{pmatrix} a\,x \\ b\,y \\ c\,z \end{pmatrix}$

DETERMINANT OF A 2×2 MATRIX

If $\mathbf{A} = \begin{pmatrix} a\,b \\ c\,d \end{pmatrix}$, det $\mathbf{A} \equiv |\mathbf{A}| = ad - bc$

INVERSE OF A 2×2 MATRIX

$$\mathbf{A}^{-1} = \frac{1}{\det \mathbf{A}} \begin{pmatrix} d & -b \\ -c & a \end{pmatrix}$$

DETERMINANT OF A 3×3 MATRIX

$$\begin{vmatrix} a_1 & b_1 & c_1 \\ a_2 & b_2 & c_2 \\ a_3 & b_3 & c_3 \end{vmatrix} = a_1 \begin{pmatrix} b_2 & c_2 \\ b_3 & c_3 \end{pmatrix} - b_1 \begin{pmatrix} a_2 & c_2 \\ a_3 & c_3 \end{pmatrix} + c_1 \begin{pmatrix} a_2 & b_2 \\ a_3 & b_3 \end{pmatrix}$$

$$= a_1 b_2 c_3 + a_2 b_3 c_1 + a_3 b_1 c_2 - a_1 b_3 c_2 - a_2 b_1 c_3 - a_3 b_2 c_1$$

INVERSE OF A 3×3 MATRIX

The cofactor of an element is found by 'removing' the row and column containing that element, and finding the determinant of the 2×2 matrix, then giving it the appropriate sign

$$+ - +$$
$$- + -$$
$$+ - +$$

Form the matrix of the cofactors, e.g. if A_1 is the cofactor of a_1, $A_1 = b_2 c_3 - b_3 c_2$.

Transpose the matrix of the cofactors, to give

$$\begin{pmatrix} A_1 & A_2 & A_3 \\ B_1 & B_2 & B_3 \\ C_1 & C_2 & C_3 \end{pmatrix} \tag{1}$$

When this matrix is multiplied by **A**, the product is a scalar multiple of the unit matrix. Divide each term in (1) by that scalar multiple to obtain the inverse of **A**. N.B. The scalar multiple is equal to det **A**.

MATRICES

A matrix is defined as a rectangular array of numbers, subject to certain rules of composition, with which we should already be familiar. A matrix with r rows and s columns is called an r by s matrix, and the rule for addition requires that if two matrices are to be added together, they must both have the same number of rows and columns, i.e. both are r by s matrices for some r, s. The rule for multiplication requires that if the product **A.B** of two matrices **A,B** exists, **A** is an r by s matrix, and **B** an s by t matrix, so that the first matrix has as many columns as the second matrix has rows.

TRANSPOSE OF A MATRIX

The transpose of a matrix **A** written $\mathbf{A}^T$ is formed by interchanging the rows and columns of **A**, i.e. if

$$\mathbf{A} = \begin{pmatrix} 1 & 2 & 3 \\ 4 & 5 & 6 \end{pmatrix}, \quad \mathbf{A}^T = \begin{pmatrix} 1 & 4 \\ 2 & 5 \\ 3 & 6 \end{pmatrix}$$

DETERMINANT OF A 2×2 MATRIX

Associated with every matrix is defined a determinant. For a 2×2 matrix, $\mathbf{A} = \begin{pmatrix} a & b \\ c & d \end{pmatrix}$, the expression $ad - bc$ is called the determinant, written det **A** or $|\mathbf{A}|$. This is useful when finding the inverse of a matrix. A matrix **A** for which det $\mathbf{A} = 0$ is called a singular matrix, and has no inverse.

INVERSE OF A MATRIX

The inverse of a matrix $\mathbf{A}$, written $\mathbf{A}^{-1}$, is the matrix such that $\mathbf{A}.\mathbf{A}^{-1}=\mathbf{A}^{-1}.\mathbf{A}=\mathbf{I}$, where $\mathbf{I}$ is the unit matrix of appropriate order. If

$$\mathbf{A}=\begin{pmatrix} a & b \\ c & d \end{pmatrix}, \mathbf{A}^{-1}=\frac{1}{\det \mathbf{A}}\begin{pmatrix} d & -b \\ -c & a \end{pmatrix}$$

GEOMETRICAL TRANSFORMATIONS

We are already familiar with geometrical transformations of a point. To find the transformation of, say, a straight line, we have to express a point on that line in parametric form. Consider, for example the transformation of the x-axis by the matrix $\begin{pmatrix} 1 & 2 \\ 3 & 4 \end{pmatrix}$. Any point on the x-axis can be written as $(X,0)$, X being a parameter. The image of this point is $\begin{pmatrix} 1 & 2 \\ 3 & 4 \end{pmatrix}\begin{pmatrix} X \\ 0 \end{pmatrix}=\begin{pmatrix} X \\ 3X \end{pmatrix}$, which always lies on the line $y=3x$, Thus the x-axis is transformed into $y=3x$

Example 19.1 Find the image of the straight line $y=2x$ under the mapping described by the matrix $\begin{pmatrix} 1 & 1 \\ -4 & 6 \end{pmatrix}$.

Any point on $y=2x$ has parametric form $(\lambda,2\lambda)$, so its image is $\begin{pmatrix} 1 & 1 \\ -4 & 6 \end{pmatrix}\begin{pmatrix} \lambda \\ 2\lambda \end{pmatrix}=\begin{pmatrix} 3\lambda \\ 8\lambda \end{pmatrix}$, which lies on the straight line $8x=3y$ for all values of λ.

MAPPING A STRAIGHT LINE ON TO ITSELF

To find the equation of a straight line that is mapped on to itself by the matrix $\begin{pmatrix} 1 & 1 \\ -4 & 6 \end{pmatrix}$, if the straight line is $y=mx$ any point on this line can be written $(\lambda,m\lambda)$ and its image will be

$$\begin{pmatrix} 1 & 1 \\ -4 & 6 \end{pmatrix}\begin{pmatrix} \lambda \\ m\lambda \end{pmatrix}=\begin{pmatrix} (1+m)\lambda \\ (-4+6m)\lambda \end{pmatrix}$$

If this lies on $y=mx$,

$$-4+6m=m(1+m)$$
$$\Rightarrow m^2-5m+4=0$$
$$\Rightarrow (m-4)(m-1)=0, m=1 \text{ or } 4$$

so each of the lines $y=x$ and $y=4x$ is mapped on to itself by this matrix. Checking, $(2,2)$ lies on $y=x$, and its image is $\begin{pmatrix} 1 & 1 \\ -4 & 6 \end{pmatrix}\begin{pmatrix} 2 \\ 2 \end{pmatrix}=\begin{pmatrix} 4 \\ 4 \end{pmatrix}$ and $(4,4)$ also lies on $y=x$; the point $(2,8)$ lies on $y=4x$, and its image under this transformation is $\begin{pmatrix} 1 & 1 \\ -4 & 6 \end{pmatrix}\begin{pmatrix} 2 \\ 8 \end{pmatrix}=\begin{pmatrix} 10 \\ 40 \end{pmatrix}$ and $(10,40)$ also lies on $y=4x$.

DETERMINANT OF A 3×3 MATRIX

The determinant of the 3×3 matrix

$$\begin{pmatrix} a_1 & b_1 & c_1 \\ a_2 & b_2 & c_2 \\ a_3 & b3 & c_3 \end{pmatrix}$$

is defined as

$$a_1\begin{vmatrix} b_2 & c_2 \\ b_3 & c_3 \end{vmatrix}-b_1\begin{vmatrix} a_2 & c_2 \\ a_3 & c_3 \end{vmatrix}+c_1\begin{vmatrix} a_2 & b_2 \\ a_3 & b_3 \end{vmatrix}$$

which, on expansion, is equal to

$$a_1b_2c_3+a_2b_3c_1+a_3b_1c_2-a_1b_3c_2-a_2b_1c_3-a_3b_2c_1$$

If preferred, the determinant can be evaluated by

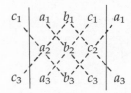

reading along the diagonals as above, signing terms $+$ along the positive diagonals and $-$ along the negative diagonals.

Example 19.2 Find the value of the determinant of the matrix $\begin{pmatrix} 1 & 2 & 3 \\ 4 & 5 & 6 \\ 7 & 8 & 9 \end{pmatrix}$

First, using the definition with 2×2 determinants,

$$1\begin{vmatrix} 5 & 6 \\ 8 & 9 \end{vmatrix}-2\begin{vmatrix} 4 & 6 \\ 7 & 9 \end{vmatrix}+3\begin{vmatrix} 4 & 5 \\ 7 & 8 \end{vmatrix}$$

$$=1\times(-3)-2\times(-6)+3\times(-3)=0$$

Secondly, using the diagonals,

$$1\times5\times9+2\times6\times7+3\times4\times8-1\times6\times8-2\times4\times9-3\times5\times7$$
$$=45+84+96-48-72-105$$
$$=0$$

MINORS AND COFACTORS

The determinant found by excluding the row and column containing a_1, i.e. $\begin{vmatrix} b_2 & c_2 \\ b_3 & c_3 \end{vmatrix}$ is called the minor of a_1; similarly the minor of b_1 is $\begin{vmatrix} a_2 & c_2 \\ a_3 & c_3 \end{vmatrix}$, of c_1 is $\begin{vmatrix} a_2 & b_2 \\ a_3 & b_3 \end{vmatrix}$.

ADJOINT MATRIX

The transpose of the matrix formed by the cofactors of the elements of a given matrix **A** is called the adjoint of **A**, i.e. if the cofactor of a_1, an element in **A**, is denoted by A_1, etc., the adjoint of **A**, written adj **A** is

$$\begin{pmatrix} A_1 & A_2 & A_3 \\ B_1 & B_2 & B_3 \\ C_1 & C_2 & C_3 \end{pmatrix}$$

INVERSE OF 3×3 MATRIX

There are many ways of finding the inverse of a 3×3 matrix, but that generally preferred at this level is given below.

Example 19.3 Find the inverse of the matrix $\mathbf{A}=\begin{pmatrix} 1 & 2 & 3 \\ 4 & 5 & 7 \\ 7 & 8 & 9 \end{pmatrix}$.

First, form the matrix of the cofactors, $\begin{pmatrix} -11 & 13 & -3 \\ 6 & -12 & 6 \\ -1 & 5 & -3 \end{pmatrix}$

where $\begin{vmatrix} 5 & 7 \\ 8 & 9 \end{vmatrix}=-11$, $\begin{vmatrix} 4 & 7 \\ 7 & 9 \end{vmatrix}=-13$, etc.

Next, transpose this matrix to obtain adj **A**, and form the product adj **A.A**, i.e.

$$\text{adj } \mathbf{A.A} = \begin{pmatrix} -11 & 6 & -1 \\ 13 & -12 & 5 \\ -3 & 6 & -3 \end{pmatrix} \begin{pmatrix} 1 & 2 & 3 \\ 4 & 5 & 7 \\ 7 & 8 & 9 \end{pmatrix} = \begin{pmatrix} 6 & 0 & 0 \\ 0 & 6 & 0 \\ 0 & 0 & 6 \end{pmatrix}$$

Since the product is $6\begin{pmatrix} 1 & 0 & 0 \\ 0 & 1 & 0 \\ 0 & 0 & 1 \end{pmatrix}$, we can deduce that $\mathbf{A}^{-1}$ is found by dividing every term in adj **A** by 6,

i.e. $\qquad \mathbf{A}^{-1} = \begin{pmatrix} -\frac{11}{6} & 1 & -\frac{1}{6} \\ \frac{13}{6} & -2 & \frac{5}{6} \\ -\frac{1}{2} & 1 & -\frac{1}{2} \end{pmatrix}$

The value of det **A** is 6, and some prefer to calculate det **A**, then divide each term of adj **A** by det **A**. But it is only too easy to make an arithmetic error when calculating the determinant of a 3×3 matrix, and as we are well advised to check by finding $\mathbf{A.A}^{-1}$ anyway, we see that this method of finding adj **A.A** (or **A**. adj **A**), which must be of

the form $\begin{pmatrix} c & 0 & 0 \\ 0 & c & 0 \\ 0 & 0 & c \end{pmatrix}$, gives us a check as well as saving the labour of

finding, and checking, the determinant of **A**.

QUESTIONS

1   Describe geometrically the transformations made on points in a plane by the following matrices

(a) $\begin{vmatrix} 3 & 0 \\ 0 & 3 \end{vmatrix}$ (b) $\begin{pmatrix} -1 & 0 \\ 0 & 1 \end{pmatrix}$ (c) $\begin{pmatrix} 0 & 1 \\ 1 & 0 \end{pmatrix}$

(d) $\begin{pmatrix} 0 & -1 \\ 1 & 0 \end{pmatrix}$ (e) $\begin{pmatrix} 0 & 1 \\ -1 & 0 \end{pmatrix}$

2   If $\mathbf{A} = \begin{pmatrix} 2 & -1 \\ 1 & 0 \end{pmatrix}$, show that **A** transforms every point on the straight line

$y = x$ into itself, and transforms every point on the straight line $y = x + 1$ into another point on the line. (Hint: take any point on $y = x + 1$ in the parametric form $(\lambda, \lambda + 1)$.)

3 Find the image of the x-axis under transformations described by each of the following matrices:

(a) $\begin{pmatrix} 1 & 3 \\ 2 & 4 \end{pmatrix}$ (b) $\begin{pmatrix} 0 & 1 \\ 2 & 3 \end{pmatrix}$ (c) $\begin{pmatrix} 1 & 1 \\ 2 & 2 \end{pmatrix}$

4 Find the equations of the straight line(s) through the origin transformed into themselves by

(a) $\begin{pmatrix} 4 & 2 \\ -1 & 1 \end{pmatrix}$ (b) $\begin{pmatrix} 7 & 1 \\ -9 & 1 \end{pmatrix}$

5 For the matrix $\begin{pmatrix} 1 & 3 & 7 \\ 2 & 5 & 9 \\ 4 & 8 & 0 \end{pmatrix}$, find

(a)   the minors and   (b)   the cofactors of 1, 3, 7 and 9.

6 Find adj $\mathbf{A}$, using $\mathbf{A}$ from Q.5.

7 Form the product adj $\mathbf{A}.\mathbf{A}$. Deduce $\mathbf{A}^{-1}$.

8 Using each of the methods given on p196, find det $\mathbf{A}$.

9 Find the inverse if it exists of each of the following matrices

(a) $\begin{pmatrix} 1 & 4 & 3 \\ -1 & 5 & 1 \\ 3 & 3 & 5 \end{pmatrix}$ (b) $\begin{pmatrix} 7 & 5 & 6 \\ 4 & 3 & 3 \\ 10 & 7 & 8 \end{pmatrix}$

10 Show that $\begin{pmatrix} 1 & 1 & -1 \\ 2 & 2 & 1 \\ -1 & -1 & -2 \end{pmatrix}\begin{pmatrix} x \\ y \\ 0 \end{pmatrix} = \begin{pmatrix} 0 \\ 0 \\ 0 \end{pmatrix}$ for all values of x and y, interpret this

geometrically.

11 If $\mathbf{A} = \begin{pmatrix} 0.8 & 0.6 \\ 0.6 & -0.8 \end{pmatrix}$, find

(a)   the image of the x-axis under this mapping,

(b)   the straight line that maps into the y-axis under $\mathbf{A}$,

(c)   the points and lines through the origin that are invariant under $\mathbf{A}$.

12 Find the value of k if the matrix $\begin{pmatrix} 1 & 2 & -1 \\ 2 & -1 & -1 \\ 0 & k & 1 \end{pmatrix}$ is singular.

13 Given that $\mathbf{A} = \begin{pmatrix} 2 & 0 & 1 \\ 0 & 1 & -1 \\ 3 & 2 & 3 \end{pmatrix}$, find $\mathbf{A}^{-1}$ and use it to obtain the solutions of

$$2x + z = 1$$
$$y - z = 4$$
$$3x + 2y + 3z = -1$$

14 Assuming matrix multiplication is associative, show that the set of all

matrices of the form $\begin{pmatrix} a & 0 & b \\ 0 & 1 & 0 \\ b & 0 & a \end{pmatrix}$ is a group under matrix multiplication.

VECTORS

CONTENTS

COMPONENTS

A vector can be expressed in terms of its components,

$$x\mathbf{i}+y\mathbf{j} \quad \text{or} \begin{pmatrix} x \\ y \end{pmatrix} \text{or occasionally } (x\ y)$$

$$x\mathbf{i}+y\mathbf{j}+z\mathbf{k} \quad \text{or} \begin{pmatrix} x \\ y \\ z \end{pmatrix} \text{or occasionally } (x\ y\ z)$$

MAGNITUDE

The magnitude of $x\mathbf{i}+y\mathbf{j}$ is $\sqrt{(x^2+y^2)}$, of $x\mathbf{i}+y\mathbf{j}+z\mathbf{k}$ is $\sqrt{(x^2+y^2+z^2)}$.

DIRECTION OF WATER

The vector $x\mathbf{i}+y\mathbf{j}+z\mathbf{k}$ makes angles α, β and γ with the coordinate axes, where

$$\cos\alpha:\cos\beta:\cos\gamma:1=x:y:z:\sqrt{(x^2+y^2+z^2)}$$

SECTION THEOREM

The position vector of the point dividing AB in the ratio $\lambda:\mu$ is $(\mu\mathbf{a}+\lambda\mathbf{b})/(\lambda+\mu)$, where $\mathbf{a},\mathbf{b}$ are the position vectors of A,B respectively.

EQUATION OF A STRAIGHT LINE

The position vector $\mathbf{r}$ of any point P on the straight line through the point position vector $\mathbf{a}$ parallel to the vector $\mathbf{b}$ is $\mathbf{r}=\mathbf{a}+t\mathbf{b}$.

SCALAR PRODUCT

By definition of the scalar product,

$$\mathbf{a}.\mathbf{b}=ab\cos\theta$$

From this, if $\mathbf{a}$, $\mathbf{b}$ are two non-zero vectors, $\mathbf{a}.\mathbf{b}=0\Longleftrightarrow\mathbf{a}$ and $\mathbf{b}$ are perpendicular.

EQUATION OF A PLANE

The equation of the plane through the point A, perpendicular to the vector **n** is **r.n=a.n**.

The equation of the plane through the points A, B, C is **r**$=(1-\lambda-\mu)$**a**$+\lambda$**b**$+\mu$**c**

COMPARISON WITH CARTESIAN COORDINATES

Compare the straight line

$$\mathbf{r}=\begin{pmatrix}a\\b\\c\end{pmatrix}+t\begin{pmatrix}p\\q\\r\end{pmatrix}$$

with
$$\frac{x-a}{p}=\frac{y-b}{q}=\frac{z-c}{r}$$

and the plane　　　$\mathbf{r}.(a\mathbf{i}+b\mathbf{j}+c\mathbf{k})=d$

with　　　$ax+by+cz=d$

MAGNITUDE AND DIRECTION OF A VECTOR

It is often convenient to describe a vector in terms of its components,

i.e. $x\mathbf{i}+y\mathbf{j}$ or $\begin{pmatrix}x\\y\end{pmatrix}$ in two dimensions, $x\mathbf{i}+y\mathbf{j}+z\mathbf{k}$ or $\begin{pmatrix}x\\y\\z\end{pmatrix}$ in three dimen-

sions. The magnitude of a vector is then $\sqrt{(x^2+y^2)}$ in two dimensions, $\sqrt{(x^2+y^2+z^2)}$ in three dimensions.

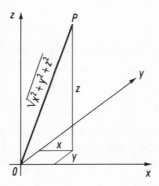

Fig 20.1

The direction of a vector in two dimensions should be found from a diagram, the angle made with the positive x-axis being arctan (y/x) when x is positive, arctan $(y/x)-\pi$ when x is negative and y positive, arctan $(y/x)-\pi$ when both are negative. *Do not try to remember these, use a diagram each time.*

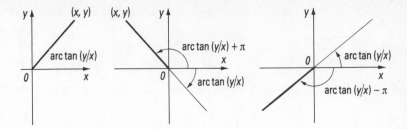

Fig 20.2

The direction of a vector in three dimensions is much harder to describe. If α, β, and γ are the angles made with the coordinate axes by the vector $\begin{pmatrix} x \\ y \\ z \end{pmatrix}$, then

$$\cos \alpha = \frac{x}{\sqrt{(x^2+y^2+z^2)}},$$

$$\cos \beta = \frac{y}{\sqrt{(x^2+y^2+z^2)}},$$

$$\cos \gamma = \frac{z}{\sqrt{(x^2+y^2+z^2)}}$$

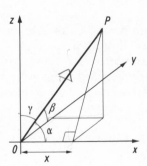

Fig 20.3

PARALLEL VECTORS AND EQUAL VECTORS

Two vectors are parallel if they are in the same direction, i.e. one must be a scalar multiple of the other. They will only be equal if every element in one is equal to the corresponding element in the other, i.e.

$$\begin{pmatrix} x \\ y \\ z \end{pmatrix} = \begin{pmatrix} a \\ b \\ c \end{pmatrix} \Leftrightarrow x=a \text{ and } y=b \text{ and } z=c$$

Example 20.1 Show that the points position vectors A $(0,1,2)$, B $(-1,0,1)$, C $(2,2,5)$ and D $(3,2,7)$ are the vertices of a trapezium.

A trapezium has one pair of opposite sides parallel. We see

$$\vec{AC}=\begin{pmatrix}2\\2\\5\end{pmatrix}-\begin{pmatrix}0\\1\\2\end{pmatrix}=\begin{pmatrix}2\\1\\3\end{pmatrix}, \quad \vec{BD}=\begin{pmatrix}3\\2\\7\end{pmatrix}-\begin{pmatrix}-1\\0\\1\end{pmatrix}=\begin{pmatrix}4\\2\\6\end{pmatrix}$$

so that $\vec{BD}=2\vec{AC}$; AC is parallel to BD and $AC=\frac{1}{2}BD$. Notice that

$$\vec{AB}=\begin{pmatrix}-1\\-1\\-1\end{pmatrix} \text{ and } \vec{CD}=\begin{pmatrix}1\\0\\2\end{pmatrix}$$

so that AB is not parallel to CD.

SECTION THEOREM

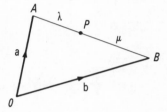

Fig 20.4

If points A,B have position vectors $\mathbf{a},\mathbf{b}$ respectively, the position vector of the point P dividing AB in the ratio $\lambda:\mu$ is $(\mu\mathbf{a}+\lambda\mathbf{b})/(\lambda+\mu)$. As a special case, the midpoint of AB has position vector $\frac{1}{2}(\mathbf{a}+\mathbf{b})$. The general result is true whether P divides AB internally or externally in the ratio $\lambda:\mu$; in the latter case, λ or μ is negative. *Always draw a diagram, and check the signs carefully.*

Example 20.2 Points A, B have position vectors

$$\mathbf{a}=\begin{pmatrix}4\\3\\5\end{pmatrix}, \quad \mathbf{b}=\begin{pmatrix}0\\-1\\1\end{pmatrix}$$

Find the position vector of (*a*) the point R dividing AB in the ratio 1:3 (*b*) the point S dividing AB in the ratio 3:1 (*c*) the point T, dividing AB externally in the ratio 3:1.

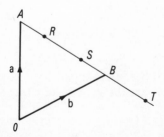

Fig 20.5

Since R divides AB in the ratio 1:3,

$$\mathbf{r}=\frac{3\mathbf{a}+\mathbf{b}}{1+3}=\tfrac{1}{4}\left[3\begin{pmatrix}4\\3\\5\end{pmatrix}+\begin{pmatrix}0\\-1\\1\end{pmatrix}\right]=\begin{pmatrix}3\\2\\4\end{pmatrix}$$

Since S divides AB in the ratio 3:1,

$$\mathbf{s}=\frac{\mathbf{a}+3\mathbf{b}}{3+1}=\tfrac{1}{4}\left[\begin{pmatrix}4\\3\\5\end{pmatrix}+3\begin{pmatrix}0\\-1\\1\end{pmatrix}\right]=\begin{pmatrix}1\\0\\2\end{pmatrix}$$

Notice that considering the first entry in each matrix, 4, 3, 1, 0 we can see that R is the point nearer to A, S the point nearer to B, as required. The point dividing AB in the ratio 1:3 is $\tfrac{1}{4}(3\mathbf{a}+\mathbf{b})$, the point dividing it in the ratio 3:1 is $\tfrac{1}{4}(\mathbf{a}+3\mathbf{b})$.

To find T, since it divides AB externally in the ratio 3:−1,

$$\mathbf{t}=\frac{1}{3-1}\left[(-1)\begin{pmatrix}4\\3\\5\end{pmatrix}+3\begin{pmatrix}0\\-1\\1\end{pmatrix}\right]=\tfrac{1}{2}\begin{pmatrix}-4\\-6\\-2\end{pmatrix}=\begin{pmatrix}-2\\-3\\-1\end{pmatrix}$$

Again, looking at Fig 20.5 shows us that we have the point we require, and that we have not inadvertently found, say, the point dividing AB in the ratio −1:3.

EQUATION OF A STRAIGHT LINE

Consider the straight line L through a fixed point A, position vector $\mathbf{a}$ parallel to a vector $\mathbf{b}$. Then any point on L can be reached from A by a translation described by $t\mathbf{b}$, for some t, since it is a translation of a variable magnitude in a specified direction. Thus the position vector $\mathbf{r}$ of any point on L is $\mathbf{r}=\mathbf{a}+t\mathbf{b}$.

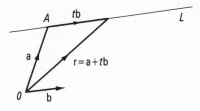

Fig 20.6

Returning to the section theorem, the point which divides the straight line through P, Q in the ratio $\lambda : \mu$ is

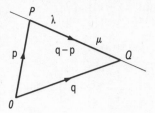

Fig 20.7

$$r = \frac{\mu p + \lambda q}{\lambda + \mu} = p + \frac{\lambda}{\lambda + \mu}(q - p)$$

$$= p + t(q - p)$$

so that, for varying t, any point on the line through A and B can be seen to have position vector

$$r = p + t(q - p)$$

as above, through the point position vector p, in the direction of the vector $(q - p)$.

Example 20.3 The vector equation of the straight line through the points position vector $\begin{pmatrix} 1 \\ 2 \\ 3 \end{pmatrix}$ parallel to the vector $\begin{pmatrix} 4 \\ 3 \\ 2 \end{pmatrix}$ is $r = \begin{pmatrix} 1 \\ 2 \\ 3 \end{pmatrix} + \lambda \begin{pmatrix} 4 \\ 3 \\ 2 \end{pmatrix}$.

Example 20.4 The vector equation of the straight line through the point position vectors $\begin{pmatrix} 2 \\ 1 \\ 4 \end{pmatrix}$ and $\begin{pmatrix} 5 \\ 2 \\ 3 \end{pmatrix}$ is $r = \begin{pmatrix} 2 \\ 1 \\ 4 \end{pmatrix} + t \begin{pmatrix} 3 \\ 1 \\ -1 \end{pmatrix}$

Taking $r = \begin{pmatrix} x \\ y \\ z \end{pmatrix}$, we see that this straight line has cartesian (parametric)

form $x = 2 + 3t$, $y = 1 + t$, $z = 4 - t$

i.e. $\quad \dfrac{x-2}{3} = \dfrac{y-1}{1} = \dfrac{z-4}{-1}$

Example 20.5 Find the position vector of the point of intersection of straight lines $r = \begin{pmatrix} 3 \\ 1 \\ 2 \end{pmatrix} + t \begin{pmatrix} 4 \\ 3 \\ 1 \end{pmatrix}$ and $r = \begin{pmatrix} 3 \\ 6 \\ 2 \end{pmatrix} + s \begin{pmatrix} -2 \\ 1 \\ 0 \end{pmatrix}$

Any point on the first line has position vector $\begin{pmatrix} 3 + 4t \\ 1 + 3t \\ 2 + t \end{pmatrix}$, any point on the second line $\begin{pmatrix} 3-2s \\ 6+\ s \\ 3 \end{pmatrix}$. At the point of intersection of the straight

lines, these vectors must be equal, so

$$3-2s=3+4t \tag{1}$$
$$1+3t=6+s \tag{2}$$
and $\quad 2+t=3 \tag{3}$

From (3), $t=1$; using (2) $s=-2$; these check in (1), so the point of

intersection of the two lines is $\begin{pmatrix} 7 \\ 4 \\ 3 \end{pmatrix}$

If the straight lines do not meet (skew lines) then the three equations in two unknowns are inconsistent, and there are no values of s and t that satisfy all three equations.

SCALAR PRODUCT

The scalar product **a.b** of two vectors **a,b** is defined as

$$\mathbf{a.b}=ab \cos \theta$$

where θ is the angle between the two vectors a,b. One particularly important consequence is that if a and b are two non-zero vectors

$$\mathbf{a.b}=0 \Leftrightarrow \mathbf{a} \text{ and } \mathbf{b} \text{ are perpendicular}$$

It is often simplest to use the **i, j, k** notation when handling scalar products, and since both this notation and the matrix notation that we have previously been using are required by most GCE Boards, we shall now use the **i, j, k** notation. Since **i, j, k** are perpendicular unit vectors,

$$\mathbf{i.j}=\mathbf{j.k}=\mathbf{k.i}=\mathbf{i.k}=\mathbf{k.j}=\mathbf{j.i}=0 \tag{1}$$
and $\quad \mathbf{i.i}=\mathbf{j.j}=\mathbf{k.k}=1 \tag{2}$

Since the scalar product is distributive over addition, we can now evaluate scalar products quite simply:

Example 20.6 If $\mathbf{a}=2\mathbf{i}+2\mathbf{j}-\mathbf{k}$, $\mathbf{b}=\mathbf{i}-\mathbf{j}+\mathbf{k}$, find **a.b**.

$$\begin{aligned}\mathbf{a.b}&=(2\mathbf{i}+2\mathbf{j}-\mathbf{k}).(\mathbf{i}-\mathbf{j}+\mathbf{k}) \\ &=2\mathbf{i.i}-2\mathbf{i.j}+2\mathbf{i.k}+2\mathbf{j.i}-2\mathbf{j.j}+2\mathbf{j.k}-\mathbf{k.i}+\mathbf{k.j}-\mathbf{k.k} \\ &=2-2-1, \text{ using (1) and (2)} \\ &=-1 \end{aligned}$$

Further, since $|\mathbf{a}|=\surd(2^2+2^2+(-1)^2)=3$ and $|\mathbf{b}|=\surd(1^2+(-1)^2+1^2)=\surd 3$

$$\mathbf{a.b}=ab \cos \theta \Rightarrow -1=3\surd 3 \cos \theta$$

so that the angle θ between the two vectors is such that
$\cos \theta = -1/(3\sqrt{3})$.

The 'angle between two straight lines' may be the acute angle or the obtuse angle, and we must check the figure carefully if we require one particular angle.

Example 20.7 Points A, B and C have position vectors $3i-2j+4k$, $7i-j+5k$ and $4i-4j+2k$ respectively. Show that AB and AC are perpendicular, and find the cosine of the angle ABC.

$AB=4i+j+k$ and $AC=i-2j-2k$, so that the scalar product is $(4i+j+k).(i-2j-2k)=0$, so the lines AB, AC are perpendicular.

To find angle ABC, we need $BC=-3i-3j-3k$

Now $|AB|=\sqrt{(4^2+1^2+1^2)}=3\sqrt{2}$
and $|BC|=\sqrt{((-3)^2+(-3)^2+(-3)^2)}=3\sqrt{3}$
so $\mathbf{a.b}=ab \cos \theta$
$$\Rightarrow (-3i-3j-3k).(-4i-j-k)=(3\sqrt{3})(3\sqrt{2}) \cos \theta$$
$$18=9\sqrt{6} \cos \theta$$
$$\cos \theta=\sqrt{\tfrac{2}{3}}$$

Notice that to ensure that we found the angle between $\overrightarrow{BA}$ and $\overrightarrow{BC}$, and not the angle between $\overrightarrow{BC}$ and $\overrightarrow{AB}$ produced, we used the vector $\overrightarrow{BA}$, not $\overrightarrow{AB}$.

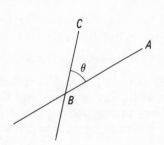

Fig 20.8

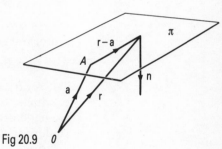

Fig 20.9

EQUATION OF A PLANE

A plane Π is fixed if we know a point through which Π passes, and a vector perpendicular to Π. If $\mathbf{a}$ is the position vector of a point in Π, $\mathbf{n}$ a vector perpendicular to Π, and $\mathbf{r}$ is the position vector of any point in Π, then $(\mathbf{r}-\mathbf{a})$ lies in Π and so is perpendicular to $\mathbf{n}$,

i.e. $(\mathbf{r}-\mathbf{a}).\mathbf{n}=0$
i.e. $\mathbf{r.n}=\mathbf{a.n}$

Example 20.8 Find the equation of the plane through the point position vector $2i-3j-4k$, perpendicular to the vector $2i-j-k$.

$$\mathbf{r.n}=\mathbf{a.n}$$
$$\Rightarrow \qquad r.(2i-j-k)=(2i-3j-4k).(2i-j-k)$$

i.e. $\qquad\qquad$ **r.(2i−j−k)=11**

This can be written $\qquad$ $2x-y-z=11$

Example 20.9 Find the position vector of the point in which the line **r=3i+2j+k+t(2i+j+2k)** meets the plane **r.(2i+3j+k)=4**

These intersect where

$$[3i+2j+k+t(2i+j+2k)].(2i+3j+k)=4$$

i.e. $\qquad\qquad\qquad\qquad\qquad 13+9t=4$

$$t=-1$$

so the position vector of their point of intersection is **i+j−k**.

EQUATION OF A PLANE THROUGH THREE POINTS

If A, B and C are three points determining a plane, the position vector relative to A of any point in that plane is $\lambda\overrightarrow{AB}+\mu\overrightarrow{AC}$, so that if the position vectors of A, B, C relative to an origin O are **a, b, c** respectively, $\overrightarrow{AB}=(\mathbf{b}-\mathbf{a})$, $\overrightarrow{AC}=(\mathbf{c}-\mathbf{a})$ and the position vector **r** of a point P relative to O is

Fig 20.10

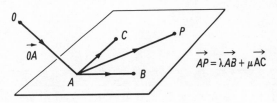

i.e. $\qquad$ $\mathbf{a}+\lambda\overrightarrow{AB}+\mu\overrightarrow{AC}$

i.e. $\qquad$ $\mathbf{a}+\lambda(\mathbf{b}-\mathbf{a})+\mu(\mathbf{c}-\mathbf{a})$,

i.e. $\qquad$ $(1-\lambda-\mu)\mathbf{a}+\lambda\mathbf{b}+\mu\mathbf{c}$

There are two parameters in this form, and it is not a form to be recommended, but it is specified at present in some GCE syllabuses.

Example 20.10 Find, in the form **r=a+sb+tc**, the vector equation of the plane through the points A, B, C, position vectors **3i+2j+k**, **i−2j−3k** and **i+3j+2k** respectively.

Since $\qquad$ $\overrightarrow{AB}=-2i-4j-4k$, $\overrightarrow{AC}=-2i+j+k$

the position vector of any point in this plane is

$$\mathbf{r}=(3i+2j+k)+s(-2i-4j-4k)+t(-2i+j+k)$$

ALTERNATIVE METHOD   A better method of finding the equation of a plane through three points is to take the form $\mathbf{r}.\mathbf{n}=k$, for some $\mathbf{n}$ and k. Then we can divide by k, and suppose that

$$\frac{1}{k}\mathbf{n}=a\mathbf{i}+b\mathbf{j}+c\mathbf{k}$$

for some a, b, c to be found, i.e. the equation of the plane is $\mathbf{r}.(a\mathbf{i}+b\mathbf{j}+c\mathbf{k})=1$. Since this passes through the point position vector $3\mathbf{i}+2\mathbf{j}+\mathbf{k}$,

$$(3\mathbf{i}+2\mathbf{j}+\mathbf{k}).(a\mathbf{i}+b\mathbf{j}+c\mathbf{k})=1$$

i.e. $3a+2b+c=1$.   (1)

Similarly, since it passes through the point $\mathbf{i}-2\mathbf{j}-3\mathbf{k}$,

$$a-2b-3c=1$$ (2)

and if it passes through $\mathbf{i}+3\mathbf{j}+2\mathbf{k}$,

$$a+3b+2c=1$$ (3)

Solving equations (1), (2) and (3) simultaneously, $a=0$, $b=1$ and $c=-1$, so the equation of the plane can be written

$$\mathbf{r}.(\mathbf{j}-\mathbf{k})=1$$

QUESTIONS

In this exercise $\mathbf{a}$, $\mathbf{b}$, $\mathbf{c}$, $\mathbf{d}$ are the position vectors of points A, B, C and D.

1 Given that

$$\mathbf{a}=\begin{pmatrix}1\\2\end{pmatrix}\quad \mathbf{b}=\begin{pmatrix}3\\7\end{pmatrix}\quad \mathbf{c}=\begin{pmatrix}6\\4\end{pmatrix}$$

(a)   show that AB is equal in length to AC;
(b)   find the vector $\mathbf{d}$ if $ABCD$ is a rhombus;
(c)   find the position vectors of M and N, the midpoints of BC and CD respectively;
(d)   find the vector $\overrightarrow{MN}$, and show $\overrightarrow{MN}=\frac{1}{2}\overrightarrow{BD}$.

2 Given

$$\mathbf{a}=\begin{pmatrix}2\\-1\end{pmatrix}\quad \mathbf{b}=\begin{pmatrix}5\\3\end{pmatrix}\quad \mathbf{c}=\begin{pmatrix}-1\\-1\end{pmatrix}$$

(a)   find $\mathbf{d}$ if $ABCD$ is a parallelogram;
(b)   find $\mathbf{d}$ if $ABDC$ is a parallelogram.

3   Given that

$$\mathbf{a}=\begin{pmatrix}-1\\-3\end{pmatrix}\ \mathbf{b}=\begin{pmatrix}2\\4\end{pmatrix}\ \mathbf{c}=\begin{pmatrix}5\\5\end{pmatrix}$$

   find   (a)   the position vector of X, the midpoint of BC;
          (b)   the position vector of Y, the midpoint of CA;
          (c)   the position vector of G, the point that divides AX in the
                   ratio 2:1;
          (d)   the position vector of G', the point that divides BY in the
                   ratio 2:1.

4   If **a**, **b**, **c** and **d** are the position vectors of points A, B, C, D respec-
     tively,

$$\text{when } \mathbf{a}=\begin{pmatrix}1\\1\\0\end{pmatrix},\ \mathbf{b}=\begin{pmatrix}3\\2\\-1\end{pmatrix},\ \mathbf{c}=\begin{pmatrix}-2\\-2\\0\end{pmatrix},\ \mathbf{d}=\begin{pmatrix}4\\1\\-3\end{pmatrix}$$

     (a)   show AB is parallel to CD
     (b)   find the lengths of AB, CD and AC

5   Points A, B, C, D have position vectors

$$\mathbf{a}=\begin{pmatrix}1\\2\\3\end{pmatrix},\ \mathbf{b}=\begin{pmatrix}-1\\-1\\2\end{pmatrix},\ \mathbf{c}=\begin{pmatrix}2\\4\\5\end{pmatrix}$$

     Find   (a)   the position vector of D, if ABCD is a parallelogram,
            (b)   the position vector of E, if ACEB is a parallelogram,
            (c)   the position vector of F, if ABCF is a trapezium with AB
                     parallel to CF and $AB=\frac{1}{2}CF$.

6   Find, in the form $\mathbf{r}=\mathbf{a}+t\mathbf{b}$, the vector equation of the line

     (a)   through A, position vector $\begin{pmatrix}2\\1\end{pmatrix}$, parallel to the vector $\begin{pmatrix}-1\\-1\end{pmatrix}$

     (b)   through A, parallel to BC, where $\mathbf{b}=\begin{pmatrix}5\\2\end{pmatrix},\ \mathbf{c}=\begin{pmatrix}-1\\3\end{pmatrix}$

     (c)   through A perpendicular to the unit vector $\begin{pmatrix}1\\0\end{pmatrix}$

7   Find which of the points A, B, C lie on the straight line

$$\mathbf{r}=\begin{pmatrix}2\\3\end{pmatrix}+t\begin{pmatrix}-1\\1\end{pmatrix}$$

     $$\text{given } \mathbf{a}=\begin{pmatrix}-1\\6\end{pmatrix}\ \mathbf{b}=\begin{pmatrix}-3\\7\end{pmatrix}\ \mathbf{c}=\begin{pmatrix}-5\\10\end{pmatrix}$$

8   Find, in vector and cartesian form, the equation of the straight lines
     AB, BC, CA given that

$$\mathbf{a}=\begin{pmatrix}1\\3\end{pmatrix}\ \mathbf{b}=\begin{pmatrix}3\\7\end{pmatrix}\ \mathbf{c}=\begin{pmatrix}5\\-1\end{pmatrix}$$

9   Find the equation of the straight line through the point position

vector $\begin{pmatrix} 3 \\ 1 \\ 2 \end{pmatrix}$ parallel to the vector $\begin{pmatrix} -1 \\ 2 \\ -3 \end{pmatrix}$

Show that this straight line passes through the point $\begin{pmatrix} 1 \\ 5 \\ -4 \end{pmatrix}$.

10   Points A, B and C have position vectors $\begin{pmatrix} 2 \\ 1 \\ 0 \end{pmatrix} \begin{pmatrix} 4 \\ 0 \\ -1 \end{pmatrix} \begin{pmatrix} 5 \\ -1 \\ 1 \end{pmatrix}$.

Find, in vector and cartesian form, the equations of the straight lines AB, BC, CA.

11   Find the scalar products $\mathbf{a}.\mathbf{b}$, $\mathbf{b}.\mathbf{c}$, $\mathbf{c}.\mathbf{a}$, where

$$\mathbf{a}=3\mathbf{i}+2\mathbf{j},\ \mathbf{b}=2\mathbf{i}-\mathbf{j}\ \text{and}\ \mathbf{c}=-2\mathbf{i}+5\mathbf{j}$$

12   Show that the vectors $2\mathbf{i}+5\mathbf{j}$ and $15\mathbf{i}-6\mathbf{j}$ are perpendicular.

13   Find the cosine of the angle between the vectors $\mathbf{i}+\mathbf{j}$ and $2\mathbf{i}-\mathbf{j}$.

14   Find the cosines of the angles between the vectors $\mathbf{a}$ and $\mathbf{b}$, $\mathbf{b}$ and $\mathbf{c}$ and $\mathbf{c}$ and $\mathbf{a}$,

$$\text{when}\ \mathbf{a}=2\mathbf{i}-3\mathbf{j},\ \mathbf{b}=3\mathbf{i}+4\mathbf{j}\ \text{and}\ \mathbf{c}=\mathbf{i}-\mathbf{j}.$$

Check that one of the angles is the sum of the other two.

15   Find a unit vector $\mathbf{u}$ in the plane of $\mathbf{i}$ and $\mathbf{j}$ perpendicular to

$$\mathbf{a}=6\mathbf{i}-8\mathbf{j}$$

by taking the unit vector $\mathbf{u}=x\mathbf{i}+y\mathbf{j}$, using $\mathbf{u}.\mathbf{a}=0$ and $|\mathbf{u}|=1$

16   Find two unit vectors perpendicular to $4\mathbf{i}-7\mathbf{j}+4\mathbf{k}$ and $2\mathbf{i}-\mathbf{j}+2\mathbf{k}$.

17   Show that the vectors $2\mathbf{i}+3\mathbf{j}+\mathbf{k}$ and $4\mathbf{i}-3\mathbf{j}+\mathbf{k}$ are perpendicular.

18   Find the cosine of the angle between the vectors $2\mathbf{i}+\mathbf{j}-\mathbf{k}$ and $\mathbf{i}-3\mathbf{j}+2\mathbf{k}$.

19   Find the cosine of the angle between the vectors $3\mathbf{j}+\mathbf{k}$, $2\mathbf{i}-\mathbf{k}$.

20   Find a unit vector in the plane of $\mathbf{i}$ and $\mathbf{j}$ perpendicular to $2\mathbf{i}-\mathbf{j}$. Find also a unit vector in three dimensions perpendicular to these vectors.

21   Find the equation of the following planes:
   (a)   through the point position vector $3\mathbf{i}-2\mathbf{j}-\mathbf{k}$ perpendicular to the vector $\mathbf{i}-2\mathbf{j}+3\mathbf{k}$
   (b)   through the point position vector $3\mathbf{i}-2\mathbf{j}+2\mathbf{k}$ perpendicular to the vector $2\mathbf{i}-3\mathbf{j}$
   (c)   through the point position vector $3\mathbf{i}+2\mathbf{j}+\mathbf{k}$ perpendicular to the vector $\mathbf{i}$.

22   Find the position vector of the point in which the line l meets the plane Π when
   (a) l is $\mathbf{r}=\mathbf{i}+t(\mathbf{j}+\mathbf{k})$ and Π is $\mathbf{r}.(3\mathbf{i}+2\mathbf{j}+\mathbf{k})=9$
   (b) l is $\mathbf{r}=2\mathbf{i}+\mathbf{j}-\mathbf{k}+t(\mathbf{i}+\mathbf{k})$ and Π is $\mathbf{r}.(4\mathbf{i}+2\mathbf{j}-3\mathbf{k})=15$
   (c) l is $\mathbf{r}=\mathbf{i}+\mathbf{j}+t\mathbf{k}$ and Π is $\mathbf{r}.(3\mathbf{i}+4\mathbf{j}+5\mathbf{k})=22$

23   (a)   Write down a vector perpendicular to the plane $x+2y+2z=6$

(b) Write down the equation of the line l through the point P position vector $3i-2j-k$ perpendicular to Π.

(c) Find the point at which l meets the plane Π.

(d) Find the perpendicular distance of P from Π.

(e) Find the position vector of the point Q, being the reflection in Π of P.

24 Find the perpendicular distance of the point P from the plane Π and the image of P in Π when

(a) $p=2i+3j+4k$, Π is $r.(2i+j-2k)=10$

(b) $p=i-2j+k$, Π is $r.(i-j)=6$

(c) $p=3i-2j-4k$, Π is $r.(3i+j+2k)=13$.

25 Find the cosine of the angle between straight lines perpendicular to each of the planes $r.(2i-j-2k)=3$ and $r.(i+j+k)=1$, and deduce the angle between the two planes.

26 Find the equation of the plane through the points position vectors $i+j+k$, $2i+j$ and $3i-j-k$,

(a) in the form $r=a+sb+tc$

(b) in the form $r.n=k$

27 Find, in the two forms given above, the equation of the plane through the points position vectors $4i-3j-2k$, $-2i+j+2k$ and $3i-j-k$.

28 Find in the two forms given above the equation of the plane through the points position vectors $i+2j+2k$, $i+j$, $j-3k$.

29 Find in the form $r.n=k$ the equation of the plane through the points position vectors $i-j+k$, j, $2i+j-4k$. Show that it passes through the point position vector $3i+3j-9k$.

30 Show that the points position vectors $-i+j$, $2i-j+k$, $3i-2j+3k$ and $i-j+4k$ lie in a plane.

31 The position vectors of points A, B and C are a, b, c respectively. Find the position vector of the fourth vertex

(a) of the parallelogram $ABCD$

(b) of the parallelogram $ABDC$.

32 Find the position vector of the point of intersection of the straight lines $r=2i+3j+k+s(i-j-k)$ and $r=i+2j+4k+t(i-2k)$, and find the cosine of the angle between these straight lines.

33 Show that the straight lines $r=3i+4j+k+s(-i+j+2k)$ and $r=i+5j+7k+t(j+k)$ are skew.

34 (a) Find the equation in a plane of the circle on points position vectors $i+j$, $3i-5j$ as diameter

(b) Find the equation in space of the sphere on the points $3i-4j+k$, $i-2j+3k$ as diameter.

35 Points A, B have position vectors a, b relative to an origin O. Show that the area of triangle $OAB=\frac{1}{2}\sqrt{[a^2b^2-(a.b)^2]}$. Hence find

(a) the area of triangle OAB, where A, B have position vectors $i-2j+3k$, $2i+2j+k$,

(b) the area of triangle ABC, C having position vector $3i+j+2k$.

36 (a) Find the position vector of the point of intersection of the line $r=3i+2j+k+t(-i+j+k)$ and the plane $r.(i+4j+3k)=2$.

(b)   Find the coordinates of the point of intersection of the line

$$\frac{x-3}{-1}=\frac{y-2}{1}=\frac{z-1}{1}$$

with the plane $x+4y+3z=2$.
Compare (a) with (b).

37   Find the cosine of the acute angle between the straight lines

$$\frac{x-1}{3}=\frac{y-3}{2}=\frac{z+2}{1} \text{ and } \frac{x+5}{2}=\frac{y-2}{-1}=\frac{z+4}{4}$$

38   Find the cosine of the acute angle between the planes

$$4x-3y+2z=1 \text{ and } 3x+2y-z=5$$

DIFFERENTIATION

CONTENTS

Learn these derivatives:

Function $F(x)$	Derivative $F'(x)$
x^n	nx^{n-1}
$\sin(ax+b)$	$a\cos(ax+b)$
$\cos(ax+b)$	$-a\sin(ax+b)$
$\tan(ax+b)$	$a\sec^2(ax+b)$
$\ln f(x)$	$\dfrac{f'(x)}{f(x)}$
$e^{f(x)}$	$f'(x)e^{f(x)}$
Product $\quad uv$	$u\dfrac{dv}{dx}+v\dfrac{du}{dx}$
Quotient $\quad \dfrac{u}{v}$	$\dfrac{v\dfrac{du}{dx}-u\dfrac{dv}{dx}}{v^2}$
Function of a function If z is a function of x, $\qquad f(z)$	$\dfrac{df}{dz}\cdot\dfrac{dz}{dx}$

PARAMETRIC FORMS

If x and y are each functions of t,

$$\frac{dy}{dx}=\frac{dy}{dt}\bigg/\frac{dx}{dt}$$

$$\frac{d^2y}{dx^2}=\frac{d}{dt}\left(\frac{dy}{dx}\right)\bigg/\frac{dx}{dt}$$

N.B. $\dfrac{dx}{dt}=1\bigg/\dfrac{dt}{dx}$ but $\dfrac{d^2x}{dt^2}$ is **not** equal to $1\bigg/\dfrac{d^2t}{dx^2}$ and $\dfrac{d^2y}{dx^2}$ is **not** equal to $1\bigg/\dfrac{d^2x}{dy^2}$.

DIFFERENTIATION OF x^n

It can be shown that if $f(x)=x^n$, the differential coefficient of $f(x)$, written $f'(x)$, is nx^{n-1}. If we write $y=x^n$, we denote the differential coefficient by dy/dx, so that if $f(x)=x^n$,

$$\frac{df}{dx} \equiv f'(x)=nx^{n-1}$$

and with the alternative notation,

$$\text{if } y=x^n, \frac{dy}{dx}=nx^{n-1}$$

Example 21.1
(a)   If $f(x)=5x^3+6x+7$, $f'(x)=15x^2+6$.
N.B. The differential coefficient of any constant, here 7, is zero.
(b)   If $f(x)=(x+1)\sqrt{x}$, $f(x)=x^{3/2}+x^{1/2}$, $f'(x)=\frac{3}{2}x^{1/2}+\frac{1}{2}x^{-1/2}$.
N.B. Remember that $\sqrt{x}=x^{1/2}$, and notice that the function can be written as the sum of two terms of the form x^n.

(c)   If $y=\dfrac{3x^3+2x+1}{x}$, $y=3x^2+2+x^{-1}$ and $\dfrac{dy}{dx}=6x-x^{-2}$.

N.B. Take care in dividing each term of the numerator by x, especially the 'easy' ones like $\dfrac{2x}{x}$ and remember $\dfrac{1}{x} \equiv x^{-1}$.

DIFFERENTIATION OF TRIGONOMETRIC FUNCTIONS

It can be shown that

$$\frac{d}{dx}(\sin x)=\cos x$$

$$\frac{d}{dx}(\cos x)=-\sin x$$

$$\frac{d}{dx}(\tan x)=\sec^2 x$$

These are the commonest of the trigonometric functions, and differential coefficients of the other functions, if required, can be deduced from these. They are

$$\frac{d}{dx}(\cot x)=-\operatorname{cosec}^2 x$$

$$\frac{d}{dx}(\sec x)=\sec x \tan x$$

and $\quad\dfrac{d}{dx}(\operatorname{cosec} x)=-\operatorname{cosec} x \cot x$

DIFFERENTIATION OF EXPONENTIAL AND LOGARITHMIC FUNCTIONS

It can be shown that

$$\frac{d}{dx}(e^x)=e^x$$

and $\quad\dfrac{d}{dx}(\ln x)\equiv\dfrac{d}{dx}(\log_e x)=\dfrac{1}{x}$

Differential coefficients of other exponential functions and of logarithms to other bases can be deduced from these.

DIFFERENTIATION OF A FUNCTION OF A FUNCTION

There are many functions that can most easily be differentiated by realizing that they are composite functions, often called functions of functions, e.g.,

$$f(x)=\sin(x^2)$$

where we have first one function of x, here x^2, then a function of that function, giving $\sin(x^2)$. Using the function notation of Chapter 1,

if $g{:}x{\rightarrow}x^2$ and $h{:}x{\rightarrow}\sin x$

$h^0g\equiv f{:}x{\rightarrow}\sin(x^2)$

Alternatively, we can write

$z=x^2,$

so $\quad f(x)=\sin z,$

and it can be shown that

$$f'(x)\equiv\frac{d}{dx}f(x)=\frac{df}{dz}\frac{dz}{dx}$$

$$=(\cos z)\,2x,\text{ since if }z=x^2,\frac{dz}{dx}=2x$$

$$=2x\cos(x^2)$$

Until we are confident, it is usually wisest to make a formal substitution as shown in these examples.

Example 21.2 If $y=\sqrt{(x^2+x+3)}$, find dy/dx.

Write $z=x^2+x+3$. Then $dz/dx=2x+1$. Since $y=\sqrt{z}\equiv z^{1/2}$,

$$\frac{dy}{dx}=\frac{dy}{dz}\frac{dz}{dx}$$
$$=(\tfrac{1}{2}z^{-1/2})(2x+1)$$
$$=\frac{2x+1}{2\sqrt{(x^2+x+3)}}$$

Example 21.3 Differentiate with respect to x, ln (sin x).

Write $z=\sin x$; then $dz/dx=\cos x$. Since $y=\ln z$,

$$\frac{dy}{dx}=\frac{dy}{dz}\frac{dz}{dx}$$
$$=\frac{1}{z}\cos x,$$
$$=\frac{\cos x}{\sin x}$$
$$=\cot x$$

Example 21.4 If $f(x)=e^{x^2+x}$, find $f'(1)$.

Here we first find $f'(x)$, then evaluate that when $x=1$.
   Write $z=x^2+x$, then $dz/dx=2x+1$. Since $f(x)=e^z$

$$f'(x)=f'(z)\frac{dz}{dx}$$
$$=e^z(2x+1)$$
$$=(2x+1)e^{x^2+x}$$
$$f'(1)=3e^2$$

PRODUCTS AND QUOTIENTS

If a function can clearly be expressed as the product of two functions, e.g. $x^2 \sin x$, then writing

$$y=x^2 \sin x$$
$$y=uv, \text{ where } u=x^2 \text{ and } v=\sin x.$$

Now $\dfrac{dy}{dx}=u\dfrac{dv}{dx}+v\dfrac{du}{dx}$

so that here, since $u=x^2$, $\dfrac{du}{dx}=2x$, and $v=\sin x$, $\dfrac{dv}{dx}=\cos x$

so $\quad \dfrac{dy}{dx}=x^2\cos x+2x\sin x$

If a function can be written as a quotient, e.g. $\dfrac{x^2}{\sin x}$ then this formula can be adapted, for if

$$y=\frac{u}{v}$$

$$\frac{dy}{dx}=u\frac{d}{dx}\left(\frac{1}{v}\right)+\frac{1}{v}\frac{du}{dx}$$

$$=u\left(-\frac{1}{v^2}\frac{dv}{dx}\right)+\frac{1}{v}\frac{du}{dx}$$

$$=\frac{v\dfrac{du}{dx}-u\dfrac{dv}{dx}}{v^2}$$

so that here, since $u=x^2$, $v=\sin x$, $\dfrac{du}{dx}=2x$ and $\dfrac{dv}{dx}=\cos x$ as before,

$$\therefore\quad \frac{dy}{dx}=\frac{(\sin x)\,2x-x^2\cos x}{(\sin x)^2}$$

$$=\frac{2x\sin x-x^2\cos x}{\sin^2 x}$$

Example 21.5 Differentiate with respect to x, $x^3\ln x$.

Write $u=x^3$, $v=\ln x$. Then $\dfrac{du}{dx}=3x^2$, $\dfrac{dv}{dx}=\dfrac{1}{x}$.

Using $\dfrac{dy}{dx}=u\dfrac{dv}{dx}+v\dfrac{du}{dx}$, the derivative is

$$x^3\left(\frac{1}{x}\right)+3x^2\ln x,$$

$$=x^2(1+3\ln x)$$

Example 21.6 Differentiate with respect to x, $\dfrac{\ln x}{x^4}$.

Write $u=\ln x$, $v=x^4$. Then $\dfrac{du}{dx}=\dfrac{1}{x}$, $\dfrac{dv}{dx}=4x^3$.

Using $\dfrac{dy}{dx} = \dfrac{v\dfrac{du}{dx} - u\dfrac{dv}{dx}}{v^2}$, the derivative is

$$= \dfrac{x^4\left(\dfrac{1}{x}\right) - (\ln x)4x^3}{(x^4)^2}$$

$$= \dfrac{1 - 4\ln x}{x^5}$$

DIFFERENTIATION OF AN IMPLICIT RELATION OR FUNCTION

Given the relation $y^2x^3 = 1$, for any positive value of x we can see that y has one of two values; the relation $y\sin x = 1$ is such that any one given value of x determines a corresponding value of y and relates y *implicitly* to x. These equations could have been rewritten $y = \pm(x^{-3/2})$ and $y = 1/\sin x$ respectively in each of which y is given *explicitly* in terms of x. Many implicit relations, however, are such that they cannot easily be rewritten to give x explicitly.

For example, if $y^2x + y\sin x = 1$, a certain given value of x will give corresponding values of y, yet it is awkward to try to 'make y the subject of the equation', i.e. to write the equation so that y is defined explicitly.

To differentiate an implicit function (or relation), remember that, since y is a function of x, $\dfrac{d}{dx}(y^2) = 2y\dfrac{dy}{dx}$, and in general

$$\dfrac{d}{dx}f(y) = \dfrac{df}{dy} \cdot \dfrac{dy}{dx}$$

e.g. $\qquad \dfrac{d}{dx}(\cos y) = -\sin y \dfrac{dy}{dx}$

and $\qquad \dfrac{d}{dx}(e^y) = e^y\dfrac{dy}{dx}$

Example 21.7 If $x^2y + y^3 = 2$, find $\dfrac{dy}{dx}$ when $x = 1$, $y = 1$.

Differentiating, $2xy + x^2\dfrac{dy}{dx} + 3y^2\dfrac{dy}{dx} = 0$,

i.e. $\qquad \dfrac{dy}{dx}(x^2 + 3y^2) = -2xy,$

$$\dfrac{dy}{dx} = \dfrac{-2xy}{(x^2 + 3y^2)}$$

When $x = 1$, $y = 1$ $\qquad \dfrac{dy}{dx} = \dfrac{-2}{4} = -\tfrac{1}{2}$

APPLICATIONS

There are several useful cases where we can change an explicit relation into an implicit relation in order to find the derivative of a function we do not already know. For example, if $y=\text{arc sin}(x/a)$, and we do not know the derivative of arc sin (x/a),

$$y=\text{arc sin}(x/a) \Rightarrow \sin y = x/a,$$

so
$$\cos y \frac{dy}{dx} = \frac{1}{a}$$

$$\frac{dy}{dx} = \frac{\frac{1}{a}}{\cos y}$$

But since $\sin y = x/a$
$$\cos^2 y = 1 - \sin^2 y$$

$$= 1 - \left(\frac{x}{a}\right)^2$$

and
$$\cos y = \pm \sqrt{\left(1 - \left(\frac{x}{a}\right)^2\right)}.$$

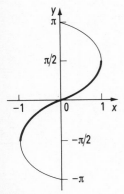

So
$$\frac{dy}{dx} = \pm \frac{\frac{1}{a}}{\sqrt{\left(1-\left(\frac{x}{a}\right)^2\right)}}$$

$$= \pm \frac{1}{\sqrt{(a^2-x^2)}}$$

Fig 21.1

Although strictly we have to consider both signs $+$ or $-$, the graph (Fig 21.1) shows that for principal values of y the gradient is positive.

We therefore usually use only that sign, and write

$$\frac{d}{dx}(\text{arc sin } x) = \frac{1}{\sqrt{(a^2-x^2)}}$$

Example 21.8 Find the derivative with respect to x of a^x.

Writing $y=a^x$, we have $\log_a y = x$. If we do not know the derivative of $\log_a y$, we write

$$\log_a y = \frac{\log_e y}{\log_e a}$$

i.e. $\log_a y = \dfrac{\ln y}{\ln a}$

i.e. $\ln y = x \ln a$

$$\frac{1}{y}\frac{dy}{dx} = \ln a$$

$$\frac{dy}{dx} = y \ln a$$

$$= a^x \ln a$$

So the derivative of a^x is $a^x \ln a$. Checking by putting $a=e$, the derivative of $e^x = e^x \ln e = e^x$, as expected.

PARAMETRIC RELATIONS

We have seen (p143) that a curve can be defined in terms of a single variable, say t, so that any one value of t determines one particular point on the curve. More generally, a relation between two variables x and y can be defined by describing each of x and y in terms of a third variable, a **parameter**, t, e.g. if $x=t^2$, $y=\sin t$, any one value of t determines both x and y. *When differentiating a relation given in terms of a parameter, do not attempt to eliminate that parameter.*

Example 21.9 If $x=t^2$, $y=t^3+t$, find dy/dx and d^2y/dx^2.

Since x and y are both functions of t,

$$\frac{dy}{dt} = \frac{dy}{dx} \cdot \frac{dx}{dt}$$

$$\therefore \qquad \frac{dy}{dx} = \frac{\dfrac{dy}{dt}}{\dfrac{dx}{dt}}$$

Since $y=t^3+t$, $dy/dt=3t^2+1$;
since $x=t^2$, $dx/dt=2t$.

So $\qquad \dfrac{dy}{dx} = \dfrac{3t^2+1}{2t}$

To find $\dfrac{d^2y}{dx^2}$, we remember that $\dfrac{d}{dt}\left(\dfrac{dy}{dx}\right) = \dfrac{d}{dx}\left(\dfrac{dy}{dx}\right) \cdot \dfrac{dx}{dt}$,

so $\qquad \dfrac{d^2y}{dx^2} = \dfrac{d}{dt}\left(\dfrac{dy}{dx}\right) \Big/ \dfrac{dx}{dt}$.

Now $\qquad \dfrac{d}{dt}\left(\dfrac{dy}{dx}\right) = \dfrac{d}{dt}\left(\dfrac{3t^2+1}{2t}\right)$

$$= \dfrac{d}{dt}\left(\dfrac{3}{2}t + \dfrac{1}{2t}\right)$$

$$= \dfrac{3}{2} - \dfrac{1}{2t^2},$$

so
$$\frac{d^2y}{dx^2}=\left(\frac{3}{2}-\frac{1}{2t^2}\right)\bigg/2t, \text{ using } \frac{dx}{dt}=2t$$

$$=\frac{3}{4t}-\frac{1}{4t^3}$$

N.B. It is true that $\dfrac{dy}{dx}=\dfrac{dy/dt}{dx/dt}$ and $\dfrac{dx}{dt}=\dfrac{1}{dt/dx}$ but remember that $\dfrac{d^2y}{dt^2}$ is

not equal to $\dfrac{d^2y}{dt^2}\bigg/\dfrac{d^2x}{dt^2}$.

QUESTIONS

1 Find dy/dx, if $y=$

(a) $\sqrt{(1+x^2)}$ (b) $\dfrac{1}{(1+2x)}$

(c) $\sin 4x$ (d) $\ln(4x)$ (e) e^{3x+2}

2 Differentiate with respect to x,

(a) $\dfrac{1}{(3x+2)}$ (b) $\sin^2 x$

(c) $\ln(\cos x)$ (d) $e^{\sin x}$ (e) $(1+\sin x)^{1/2}$

3 (a)   If $f(x)=\ln(x^2+2x+3)$, find $f'(0)$

    (b)   If $f(x)=\dfrac{1}{1+\sin x}$, find $f'(\pi/6)$.

    (c)   If $f(x)=e^{\tan x}$, find $f'(0)$.

Differentiate with respect to x:

4 $x\sin 3x$

5 $x^2\cos 2x$

6 $x^3\tan x$

7 xe^{2x}

8 x^2e^{2x+1}

9 $x\ln x$

10 $x^2(x+1)^{\frac{1}{2}}$

11 $\dfrac{\sin x}{x}$

12 $\dfrac{\cos x}{x^2}$

13 $\dfrac{\sin x}{e^x}$

14 $\dfrac{\ln x}{x^2}$

15 $\dfrac{\cos 2x}{\sin x}$

16 Find dy/dx in each of the following:

(a) $y+x\sin y=1$ (b) $xy+y^3=\sin x$ (c) $ye^x+y^2=x^2$

17   If $xy^2+yx^2=2$, find dy/dx at the point $(1,1)$.

18   If $y=\arccos(\sqrt{x})$, show that $dy/dx=-1/\sin 2y$.

19   If $y=x^x$, show that $dy/dx=y(1+\ln x)$.

20   Find dy/dx if
   (a) $x=t^2,\ y=t^3$,
   (b) $x=\sin 2t,\ y=\cos t$,
   (c) $x=\sin^2 t,\ y=\cos^2 t'$.

21   If $x=2t/(1-t^2)$ and $y=(1+t^2)/(1-t^2)$ find d^2y/dx^2 in terms of t.

22   Find dy/dx if $y=\ln(+\cos 2x)$.

23   Differentiate with respect to x,
   (a) $\ln(x^2)$
   (b) $(\ln x)^2$

24   If $f(x)=(1-x^2)^{1/2}\arcsin x$, find $f'(x)$ in its simplest form.

25   A curve is given parametrically by $x=t+\sin t$, $y=1+\cos t$. Find dy/dx, and the equation of the tangent at the point where $t=\pi/2$.

26   Differentiate with respect to x,

   (a) $\dfrac{x^2-1}{x^2+1}$ (b) $\dfrac{x^2-2}{x+1}$

27   Find the derivative with respect of x of $\ln\left(\dfrac{x}{1+x^2}\right)$

28   If $ye^x=(1+x)^2$, find dy/dx in terms of x.

29   If $y=\ln\sin x$, prove that $\dfrac{d^2y}{dx^2}+\left(\dfrac{dy}{dx}\right)^2+1=0$.

30   If $\dfrac{1}{x}+\dfrac{1}{y}-\dfrac{1}{a}=0$, show that $\dfrac{dy}{dx}=-\dfrac{y^2}{x^2}$.

31   If $y=\tan^2 x$, show that $\dfrac{d^2y}{dx^2}=6y^2+8y+2$.

APPLICATIONS OF DIFFERENTIAL CALCULUS

CONTENTS

NOTES

MAXIMA AND MINIMA

At a **maximum** of $f(x)$, $f'(x)=0$ and is decreasing;
at a **minimum of** $f(x)$, $f'(x)=0$ and is increasing;
at a **point of inflexion**, $f''(x)=0$ and changes sign; $f'(x)$ may or may not equal zero.

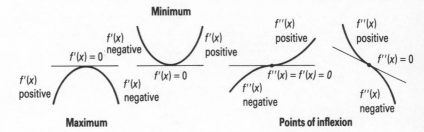

Fig 22.1

Maxima and minima are sometimes called **turning points**.

All points for which $f'(x)=0$ are called **stationary points**, and the corresponding values of $f(x)$ are called **stationary values**. Thus maxima and minima are stationary points, and so are some points of inflexion.

If $f(x)$ has a stationary value at $x=a$,

$$f'(x)=(x-a)F(x);$$

if $(x-a)$ is not a factor of $F(x)$, $x=a$ will be a maximum if $F(a)$ is negative, a minimum if $F(a)$ is positive.

RATES OF CHANGE

If y is a function of x, and x is a function of t,

$$\frac{dy}{dt}=\frac{dy}{dx}\cdot\frac{dx}{dt}$$

SMALL INCREMENTS

If y is a function of x, then a small increment δx in x produces a corresponding small increment δy in y, where

$$\delta y\approx\frac{dy}{dx}\,\delta x$$

MAXIMA AND MINIMA

A function y of x is said to have a maximum when $x=x_1$ if the value of y when $x=x_1$ is greater than the value of y at points immediately on either side of it, i.e. $f(x_1)>f(x)$ for all values of x'near to' x_1. Similarly, the function is said to have a minimum at $x=x_2$ if

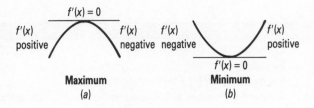

Fig 22.2

$f(x_2)<f(x)$ for all values of x'near to' x_2. Note that the maxima and minima are not necessarily the greatest and least values in a range (Fig 22.3).

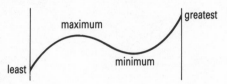

Fig 22.3

From Fig 22.2(a), we see that at $x=x_1$, $f'(x_1)=0$, and that the sign of the gradient changes from positive to negative at a maximum, whereas at a minimum the sign changes from negative to positive; this gives us a good way of distinguishing between maxima and minima. A point at which the gradient is zero, but the sign of the gradient does not change, is called a **point of inflexion**.

Example 22.1 Find the values of x which gives stationary values of the function $f(x)=x^3(x+1)^2$, distinguishing between maxima, minima and points of inflexion.

All points for which $f'(x)=0$ are called **stationary points**, and the corresponding values of $f(x)$ are called **stationary values**. These include maxima, minima and points of inflexion.

Since
$$f(x)=x^3(x+1)^2$$
$$f'(x)=3x^2(x+1)^2+x^3[2(x+1)]$$
$$=x^2(x+1)[3(x+1)+2x]$$
$$=x^2(x+1)(5x+3)$$

so that $f'(x)=0$ when $x=0$, -1 or -0.6.

To investigate the nature of the stationary point when $x=-1$, write

$$f'(x)=(x+1)x^2(5x+3).$$

Then when x is slightly less than -1, say -1.1, $f'(x)$ is the product of three terms one negative, one positive, and one negative, and so $f'(x)$ must be positive.

i.e., since $f'(x)=(x+1)x^2(5x+3)$
for x slightly less than -1 $f'(x)=(\text{negative})(\text{positive})(\text{negative})$

When x is slightly more than -1, e.g. $x=-0.9$,

$$f'(x)=(\text{positive})(\text{positive})(\text{negative})$$

since $x+1$ is now positive, so that $f'(x)$ is negative. Thus around $x=-1$, $f'(x)$ is positive, then zero, and then negative, so that $x=-1$ is a maximum.

To investigate $x=-0.6$, when x is slightly less than -0.6, say -0.7,

since $f'(x)=(x+1)x^2(5x+3)$
                          $f'(x)=(\text{positive})(\text{positive})(\text{negative})$

and when x is slightly more than -0.6, say -0.5,

$$f'(x)=(\text{positive})(\text{positive})(\text{positive})$$

so that around $x=-0.6$, $f'(x)$ is negative, then zero, and then positive. Thus $x=-0.6$ is a maximum.

To investigate $x=0$, near $x=0$ both $x+1$ and $5x+3$ are positive. x^2 does not change sign, being a square, so that $f'(x)$ does not change sign, and $f'(x)$ is positive, then zero, and then positive. Thus $x=0$ is a point of inflexion.

The nature of these stationary points is shown in Fig 22.4.

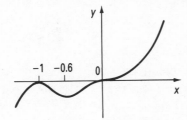

Fig 22.4

The method can be abbreviated thus:
If $f'(x)$ has a stationary value at $x=a$,

$$f'(x)=(x-a)F(x)$$

If $(x-a)$ is not a factor of $F(x)$, $F(x)$ does not change sign as x increases through $x=a$, so that if $F(a)$ is positive, $f'(x)$ is

          negative   zero   positive

and $x=a$ is a minimum; if $F(a)$ is negative, the sign of $f'(x)$ is

          positive   zero   negative

and $x=a$ is a maximum. If, however, $(x-a)$ is a factor of $F(x)$, this abbreviation needs modifying.

Look at Example 22.1. Considering the stationary value $x=-1$,

$$f'(x)=(x+1)x^2(5x+3)$$

Here, $F(x)=x^2(5x+3)$, and $F(-1)=-2$, negative, so $x=-1$ is a maximum. Considering the stationary value $x=-0.6$,

$$f'(x)=(5x+3)x^2(x+1)$$

Now $F(x)=x^2(x+1)$ and $F(-0.6)$ is positive, so $x=-0.6$ is a minimum. When we consider $x=0$,

$$f'(x)=x^2(x+1)(5x+3)$$

If we take $F(x)=(x+1)(5x+3)$, $F(0)=3$, positive. Since x^2 is positive on either side of $x=0$, $f'(x)$ is always positive near $x=0$, and we have a point of inflexion.

N.B. Write $f'(x)=(x-a)F(x)$, **not** $f'(x)=(a-x)F(x)$.

USE OF SECOND DERIVATIVE

At a maximum, we have seen that $f'(x)$ is positive, then zero, and then negative, and so seems to be decreasing with its own second derivative, $f''(x)$, being negative. Similarly, it seems that at a minimum $f''(x)$ is positive. Indeed, both conclusions will often be true, but the next example shows that there are exceptions.

Example 22.2 Find the nature of the stationary point of the function $f(x)=(x-1)^4$.

Since $f(x)=(x-1)^4$

$$f'(x)=4(x-1)^3,$$

so $f'(x)=0 \Rightarrow 4(x-1)^3=0$, i.e. $x=1$.

When x is slightly less than 1, $f'(x)$ is negative; when x is slightly more than 1, $f'(x)$ is positive, so that $f'(x)$ is negative, then zero, then positive, and $x=1$ is seen to be a minimum.

Notice that in this case $f''(x)=12(x-1)^2$, so $f''(1)=0$, and we cannot identify this minimum by using the test of the second derivative.

SUMMARY

These findings can be summarized:

at a **maximum**, $f'(x)=0$ and $f''(x)$ is not positive;
at a **minimum**, $f'(x)=0$ and $f''(x)$ is not negative;
at a **point of inflexion**, $f''(x)=0$; *note that $f'(x)$ may or may not be zero.*

if $f'(x_1)=0$ and $f''(x_1)<0$, $x=x_1$ is a maximum;
if $f'(x_2)=0$ and $f''(x_2)>0$, $x=x_2$ is a minimum;

if $f''(x_3)=0$ and $f''(x_3)$ changes sign as x increases through the value $x=x_3$, then $x=x_3$ is a point of inflexion (Fig 22.5).

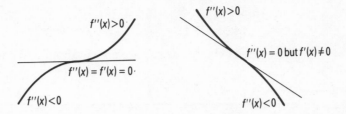

Fig 22.5

MAXIMUM AND MINIMUM VALUES

The function $f(x)$ has a maximum when $x=x_1$ if $f(x_1)>f(x)$ when x is 'near to' x_1; the value of $f(x_1)$ is called the **maximum value**. Similarly, if $x=x_2$ is a minimum, then $f(x_2)$ is the **minimum** value. Distinguish carefully whether an examination question requires the values of x that give maxima or minima, i.e. x_1 or x_2, or whether it requires the maximum or minimum values, i.e. $f(x_1)$ or $f(x_2)$.

RATES OF CHANGE

The radius r of a sphere determines the volume V of the sphere, for V is a function of r. If r varies with time t, then,

since $\quad V=\dfrac{4}{3}\pi r^3,$

$$\frac{dV}{dt}=4\pi r^2\frac{dr}{dt}$$

Knowing r and dr/dt, we can find dV/dt.

Example 22.3 If V and S denote the volume and surface area respectively of a sphere radius r, find dV/dt and dS/dt when $r=5$ and $dr/dt=0.2$, the units being cm and seconds.

Since $\quad V=\dfrac{4}{3}\pi r^3,$

$$\frac{dV}{dt}=4\pi r^2\frac{dr}{dt}$$

$$=4\pi(5)^2(0.2)$$
$$\approx62.8$$

The rate of change of volume is about $62.8\,\mathrm{cm}^3\,\mathrm{s}^{-1}$.

Since $S = 4\pi r^2$,

$$\frac{dS}{dt} = 8\pi r \frac{dr}{dt}$$

$$\approx 25.1$$

The rate of change of surface area is about $25.1 \, \text{cm}^2 \, \text{s}^{-1}$.

SMALL INCREMENTS

Figure 22.6 illustrates that if δy is a small change in y produced by a small change δx in x, then

$$\delta y \simeq \frac{dy}{dx} \delta x$$

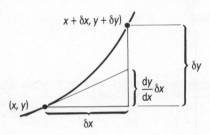

Fig 22.6

The application of this result is very similar to that of rates of change.

Example 22.4 Find the change in volume of a sphere, radius r, when the radius increases from 5 cm to 5.1 cm.

$$\text{Since } V = \frac{4}{3} \pi r^3$$

$$\delta V \simeq 4\pi r^2 \delta r$$

so the change in V is approximately $4\pi(5)^2(0.1)$, about $31.4 \, \text{cm}^3$.

Example 22.5 The time T of oscillation of a pendulum length l is $2\pi\sqrt{(l/g)}$. Find the percentage change in T if l increases from 0.5 m to 0.52 m, g remaining constant.

The formula can be written $T = cl^{1/2}$

so $\delta T \simeq c\tfrac{1}{2}l^{-1/2}\delta l$

and $\dfrac{\delta T}{T} \times 100 = \dfrac{c\tfrac{1}{2}l^{-1/2}}{cl^{1/2}} \, \delta l \times 100$

$$= \frac{\delta l}{2l} \times 100$$

$$= \frac{0.02}{2 \times 0.5} \times 100$$

$$= 2$$

So there is approximately a 2% change in T for this change in l.

QUESTIONS

1   Find the nature of the turning points of

    (a) $y = x^3$
    (b) $y = (x-1)^2(x+1)^2$
    (c) $y = x^3(3x+1)$
    (d) $y = x\sqrt{(1+x)}$

2   Find the maxima and minima of the following functions, and the maximum and minimum values of the functions at those points:

    (a) $y = (x-1)^4 + 3$
    (b) $y = x^2(x-1)$
    (c) $y = \sin^2 x \cos x$ for $0 < x < \pi/2$
    (d) $y = xe^{-\frac{1}{2}x^2}$

3   If x and y are both functions of t, find $\dfrac{dy}{dt}$ if

    (a) $y = x^4$, $x = 5$ and $dx/dt = 0.2$.
    (b) $y = \sin^2 x$, $x = \pi/4$ and $dx/dt = 5$.
    (c) $y = \ln \sin x$, $x = \pi/6$ and $dx/dt = 4$.

4   Find the change in the surface area of a sphere when the radius increases from $10\,\text{cm}$ to $10.1\,\text{cm}$.

5   Find the percentage change in the volume of a sphere when the radius increases from $8\,\text{cm}$ to $8.1\,\text{cm}$. Compare this approximation with the exact percentage change, by calculating the volume of spheres radius $8\,\text{cm}$ and $8.1\,\text{cm}$.

6   The surface area of a solid cylinder in which the height is twice the base-radius r is $6\pi r^2$. Find the percentage change in the surface area when the radius increases from $5\,\text{cm}$ to $5.1\,\text{cm}$.

7   Find the nature of the turning points of the curve $y = \dfrac{x}{1+x^2}$.

8   Find the nature of the turning points of the curve $y = (1+x)^2 e^{-2x}$.

9   Find the maximum value of $f(t) = 2e^{-t} - 3e^{-2t}$. If t is restricted so that $0 \leqslant t \leqslant 1$, find the greatest value of $f(t)$.

10 Given that x and y vary so that $3x + 4y = 5$, write $x^2 + y^2$ in terms of x only, and hence show that the minimum value of $x^2 + y^2$ is 1. Interpret this result geometrically.

11 The area of a triangle is to be calculated from measurements using the fomula $\triangle = \frac{1}{2}bc \sin A$. Find the percentage error in calculating the area if

    (a) b and A are measured accurately, but there is an error of 1% in measuring c, (*b*)   *b* and *c* are measured accurately, but *A* is measured as 1.1 rad instead of 1 rad.

12   Points *A* and *B* lie on a circle centre O, radius *a*. If angle AOB is $2x$, the radius of the inscribed circle of triangle AOB is $a \tan x(1 - \sin x)$. Show that this radius has a maximum value when $\sin x = \frac{1}{2}(\sqrt{5} - 1)$.

13   A wire of length $2a$ is bent to form the sides of an isosceles triangle. Denoting the lengths of the sides by x, x and $2a - 2x$, show that the area of the triangle is $(a - x)\sqrt{(2ax - a^2)}$ and that this area is a maximum when the triangle is equilateral.

14   Find the coordinates of the turning points of the function

$$f(x) = \frac{\cos x}{2 - \sin x}$$ in the interval $0 < x < 2\pi$, distinguishing between

maxima and minima.

15   If $y = x^2 + \dfrac{8}{1 - x}$, show that the only turning point occurs at $x = -1$, and

that there is only one point of inflexion.

16   A particle moves along the *x*-axis so that its displacement *x* metres from the origin O at time *t* seconds is given by $x = \sin 2t - t$. Find the greatest distance of the particle from the origin if $0 < t < \pi$.

INTEGRATION

CONTENTS

NOTES

The following integrals must be **learnt**:

$F(x)$	$\int F(x)dx$
$x^n, n \neq -1$	$\dfrac{x^{n+1}}{n+1}+C$
$\sin (ax+b)$	$-\dfrac{1}{a}\cos (ax+b)+C$
$\cos (ax+b)$	$\dfrac{1}{a}\sin (ax+b)+C$
$\dfrac{1}{ax+b}$	$\dfrac{1}{a}\ln(ax+b)+C$
$\dfrac{f'(x)}{f(x)}$	$\ln f(x)+C$
e^{ax}	$\dfrac{1}{a}e^{ax}+C$
$\dfrac{1}{\sqrt{(a^2-x^2)}}$	$\arcsin \dfrac{x}{a}+C$
$\dfrac{1}{a^2+x^2}$	$\dfrac{1}{a}\arctan \dfrac{x}{a}+C$

Trigonometric identities are often useful, e.g.

$$\int \sin^2 \theta \, d\theta = \int \tfrac{1}{2}(1-\cos 2\theta)\, d\theta$$
$$= \tfrac{1}{2}\theta - \tfrac{1}{4}\sin 2\theta + C$$
$$\int \sin 3\theta \cos \theta \, d\theta = \int \tfrac{1}{2}(\sin 4\theta + \sin 2\theta)\, d\theta$$
$$= -\tfrac{1}{8}\cos 4\theta - \tfrac{1}{4}\cos 2\theta + C$$

So are partial fractions, e.g.

$$\int \frac{dx}{x(x-1)} = \int \left(\frac{-1}{x} + \frac{1}{(x-1)}\right) dx$$

$$= \ln \left| \frac{x-1}{x} \right| + C$$

If there is a product, see whether one factor is the derivative of part of the other, e.g.

$\int (2x+1)(x^2+x+1)^{\frac{1}{2}} dx = \frac{2}{3}(x^2+x+1)^{\frac{3}{2}}+C$

otherwise you may need to use **integration by parts**, e.g.

$$\int u\frac{dv}{dx}dx = uv - \int v\frac{du}{dx}dx$$

AREAS AND VOLUMES

If the region R_1 is bounded by the x-axis, the curve $y=f(x)$ and the lines $x=a$, $x=b$, the **area** of R_1 is $\int_a^b y\ dx$; the **volume of the solid** formed when R_1 is rotated completely about the x-axis is $\pi \int_a^b y^2\ dx$.

If the region R_2 is bounded by the y-axis, the curve $x=F(y)$ and the lines $y=c$, $y=d$, the area of R_2 is $\int_c^d x\ dy$; the volume of the solid formed when R_2 is rotated completely about the y-axis is $\pi \int_c^d x^2\ dy$.

MEAN VALUE

If $y=f(x)$, the mean value $\bar{y}$ of y over the interval $a\leqslant x\leqslant b$ is

$$\frac{1}{b-a}\int_a^b y\ dx$$

We must be thoroughly familiar with the integrals listed above, even if we are taking an examination in which formula sheets are provided.

TRIGONOMETRIC INTEGRALS

Many trigonometric functions can be integrated using the formulae on p173, expressing a product of two functions as the sum or difference of two functions. All of these will be fairly simple, and should avoid any very awkward algebra.

Example 23.1

(a)
$$\sin^2 2\theta\,d\theta = \int \tfrac{1}{2}(1-\cos 4\theta)d\theta$$
$$= \tfrac{1}{2}\theta - \tfrac{1}{8}\sin 4\theta + C$$

(b)
$$\cos 4\theta \cos \theta\,d\theta = \int \tfrac{1}{2}(\cos 5\theta + \cos 3\theta)\,d\theta$$
$$= \tfrac{1}{10}\sin 5\theta + \tfrac{1}{6}\sin 3\theta + C$$

But (c) $\int \sin^7 \theta \cos \theta\,d\theta = \tfrac{1}{8}\sin^8 \theta + C,$

noticing that $\cos \theta$ is the derivative of $\sin \theta$. Any attempt to use sums and products formulae here soon leads to heavy algebra.

LOGARITHMIC INTEGRALS

To integrate a quotient, always look to see if the numerator is the derivative of the denominator, or if by division it can be rearranged so that it is.

Example 23.2

(a) $$\int \frac{x^2}{x^3+5}dx = \int \frac{\frac{1}{3}(3x^2)}{x^3+5}dx = \frac{1}{3}\ln(x^3+5)+C$$

(b) $$\int \frac{\cos x}{1+\sin x}dx = \ln(1+\sin x)+C$$

(c) $$\int \frac{x}{x+1}dx = \int \left(\frac{x+1}{x+1}-\frac{1}{x+1}\right)dx$$
$$= x - \ln(x+1)+C$$

(d) $$\int \frac{1}{1+e^x}dx = \int \left(\frac{1+e^x}{1+e^x}-\frac{e^x}{1+e^x}\right)dx$$
$$= x - \ln(1+e^x)+C$$

Strictly, of course, $\int \frac{1}{x}dx = \ln|x|+C$, but the integrand $\frac{1}{x}$ is usually positive over the interval chosen so that we can generally omit the modulus sign.

PARTIAL FRACTIONS

If the integrand contains a product in the denominator, then we usually need to express it in partial fractions.

Examples 23.3

(a) $$\int \frac{1}{(x-1)(x+2)}dx = \int \left(\frac{\frac{1}{3}}{x-1}-\frac{\frac{1}{3}}{x+2}\right)dx$$
$$= \frac{1}{3}\ln\left(\frac{x-1}{x+2}\right)+C$$

(b) $$\int_1^2 \frac{x-1}{(x+1)(x^2+1)}dx = \int \left(\frac{x}{x^2+1}-\frac{1}{(x+1)}\right)dx$$
$$= [\frac{1}{2}\ln(x^2+1)-\ln(x+1)]_1^2$$
$$= \frac{1}{2}\ln 5 - \ln 3 - (\frac{1}{2}\ln 2 - \ln 2)$$
$$= \frac{1}{2}\ln (10/9)$$

INVERSE TRIGONOMETRIC INTEGRALS

Some examination syllabuses require that we are familiar with

$$\int \frac{1}{\sqrt{(a^2-x^2)}}dx = \arcsin\left(\frac{x}{a}\right) \text{ and } \int \frac{1}{a^2+x^2}dx = \frac{1}{a}\arctan\left(\frac{x}{a}\right).$$

The integrals that are met at this stage usually only require a given integral to be compared with a standard form.

Example 23.4

(a) $\displaystyle\int \frac{1}{\sqrt{(9-x^2)}}dx = \arcsin(x/3)+C$

(b) $\displaystyle\int \frac{1}{\sqrt{(9-4x^2)}}dx = \frac{1}{2}\int \frac{1}{\sqrt{(\frac{9}{4}-x^2)}}dx$

                                  $= \frac{1}{2}\arcsin(2x/3)+C$

(c) $\displaystyle\int \frac{1}{9+4x^2}dx = \frac{1}{4}\int \frac{1}{(3/2)^2+x^2}dx$

                           $= \frac{1}{4}(\frac{2}{3})\arctan\left(\frac{x}{3/2}\right)+C$

                           $= \frac{1}{6}\arctan\left(\frac{2x}{3}\right)+C$

Very occasionally we may need to 'complete the square' in the denominator and we may need to make a substitution, as in the next section.

INTEGRALS REQUIRING SUBSTITUTIONS

To find $\displaystyle\int \frac{1}{x^2+2x+2}dx$, we write the denominator as $(x+1)^2+1$.

We can now substitute $z=x+1$, i.e. $dx/dz=1$, so that

$$\int \frac{1}{(x+1)^2+1}dx = \int \frac{1}{z^2+1}\left(\frac{dx}{dz}\right)dz$$

$$= \arctan z+C \left(\text{since}\frac{dx}{dz}=1\right)$$

$$= \arctan(x+1)+C$$

Notice that the original integral $\int dx$ must be changed into $\displaystyle\int \frac{dx}{dz}dz$.

We must take care always to write down the 'dx', otherwise it is easy to forget that the second integral must be $\int dz$. You can expect that all substitutions except the very simplest will be given in an examination.

Example 23.5 Find $\int \surd(a^2-x^2)dx$, using the substitution $x=a \sin \theta$.

If $x=a \sin \theta$, $\dfrac{dx}{d\theta}= a \cos \theta$, so

$$\int \surd(a^2-x^2)dx=\int \surd(a^2-x^2)\frac{dx}{d\theta}d\theta$$

$$=\int \surd(a^2-a^2 \sin^2\theta)(a \cos \theta)\,d\theta$$
$$=a^2\int \cos^2\theta\,d\theta$$
$$=\tfrac{1}{2}a^2\int(1+\cos 2\theta)\,d\theta$$
$$=\tfrac{1}{2}a^2[\theta+\tfrac{1}{2}\sin 2\theta]+C$$
$$=\tfrac{1}{2}a^2[\theta+\sin \theta \cos \theta]+C$$

$$=\tfrac{1}{2}a^2\left[\arcsin (x/a)+\left(\frac{x}{a}\right)\surd\left(1-\frac{x^2}{a^2}\right)\right]+C$$

$$=\tfrac{1}{2}a^2 \arcsin (x/a)+\tfrac{1}{2}x\surd(a^2-x^2)+C$$

When we have a definite integral, it is almost invariably best to change the limits of the integral to avoid substituting at the end. Here

$$\int_{x=0}^{x=\frac{1}{2}a} \surd(a^2-x^2)\,dx=a^2\int_{\theta=0}^{\theta=\pi/6} \cos^2 \theta\,d\theta,$$

since when $x=\tfrac{1}{2}a,$ $\tfrac{1}{2}a=a \sin \theta \Rightarrow\theta=\pi/6$
and when $x=0,$ $0=a \sin \theta \Rightarrow\theta=0$

$$a^2\int_0^{\pi/6} \cos^2 \theta\,d\theta=\tfrac{1}{2}a^2\left[\theta+\sin \theta \cos \theta\right]_0^{\pi/6}, \text{ as before,}$$

$$=\tfrac{1}{2}a^2\left[\frac{\pi}{6}+\tfrac{1}{4}\surd 3\right]$$

INTEGRATION BY PARTS

Since $\dfrac{d}{dx}(uv)=v\dfrac{du}{dx}+u\dfrac{dv}{dx}$

$$\int u\frac{dv}{dx}\,dx=uv-\int v\frac{du}{dx}\,dx \qquad\qquad (1)$$

Often we can differentiate one term of a product and make it simpler, while when we integrate the other it does not become much more difficult. For example, considering $\int x \ln x\,dx$, we see that $\dfrac{d}{dx}(\ln x)=\dfrac{1}{x}$, whereas $\int x\,dx=\tfrac{1}{2}x^2$, so that we *may* obtain an integral that is simpler than our initial integral.

Set out clearly which function we are differentiating, here $u=\ln x$, $\dfrac{du}{dx}=\dfrac{1}{x}$, and which we are integrating, here $\dfrac{dv}{dx}=x$, $v=\tfrac{1}{2}x^2$, and substitute in (1),

$$\int x \ln x \, dx = \tfrac{1}{2}x^2 \ln x - \int (\tfrac{1}{2}x^2)\frac{1}{x} dx$$

$$= \tfrac{1}{2}x^2 \ln x - \int \tfrac{1}{2}x \, dx$$

$$= \tfrac{1}{2}x^2 \ln x - \tfrac{1}{4}x^2 + C$$

Example 23.6 Find $\int x \arctan x \, dx$

Since we do not know $\int \arctan x \, dx$, we cannot integrate that, and anyway $\frac{d}{dx}(\arctan x) = \frac{1}{1+x^2}$, which may seem a little simpler, so integrate x and differentiate $\arctan x$. Thus if

$$u = \arctan x, \ \frac{du}{dx} = \frac{1}{1+x^2}$$

and if $\quad \frac{dv}{dx} = x, \ v = \tfrac{1}{2}x^2$

so $\quad \int x \arctan x \, dx = \tfrac{1}{2}x^2 \arctan x - \int \frac{\tfrac{1}{2}x^2}{1+x^2} dx$

Now $\quad \int \frac{x^2}{1+x^2} dx = \int \frac{x^2 + 1 - 1}{1+x^2} dx$

$$= \int \left(1 - \frac{1}{1+x^2}\right) dx$$

$$= x - \arctan x$$

so $\quad \int x \arctan x \, dx = \tfrac{1}{2}x^2 \arctan x - \tfrac{1}{2}(x - \arctan x) + C$

$$= \tfrac{1}{2}(x^2 + 1) \arctan x - \tfrac{1}{2}x + C$$

Sometimes it is useful to consider a function $f(x)$ as $1 \times f(x)$, in order to differentiate $f(x)$ and integrate the 1.

Example 23.7 Find $\int \ln x \, dx$.

Write $\quad \int \ln x \, dx = \int 1 \times \ln x \, dx$.

If $\quad u = \ln x, \frac{du}{dx} = \frac{1}{x}$ and if $\frac{dv}{dx} = 1, \ v = x,$

so $\quad \int \ln x \, dx = x \ln x - \int x \frac{1}{x} dx$

$$= x \ln x - x + C$$

AREAS OF REGIONS

The area of the region bounded by the curve $y=f(x)$, straight lines $x=a$, $x=b$ and the x-axis is $\int_a^b y\,dx$ (Fig 23.1(a)); the region bounded by the curve $x=F(y)$, the lines $y=c$, $y=d$ and the y-axis is $\int_c^d x\,dy$ (Fig 23.1(b)).

Take care that the integrand is positive throughout the interval, and where possible, always check approximately the answer.

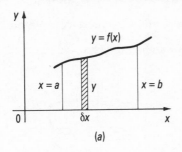
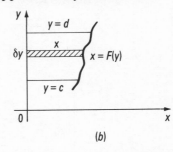

Fig 23.1

(a) (b)

Example 23.8 Find the area of the region bounded by the curves $y=x^2$ and $x=y^2$.

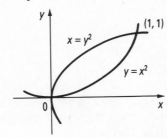

Fig 23.2

Solving simultaneously we see (Fig 23.2) that the curves meet at $(0,0)$ and at $(1,1)$. The area of the region bounded by $x=y^2$, $x=1$ and the x-axis is

$$\int_0^1 y\,dx=\int_0^1 x^{1/2}\,dx$$

$$=\left[\tfrac{2}{3}x^{3/2}\right]_0^1$$

$$=\tfrac{2}{3}$$

The area of the region bounded by $y=x^2$, $x=1$ and the x-axis is

$$\int_0^1 x^2\,dx=\left[\tfrac{1}{3}x^3\right]_0^1$$

$$=\tfrac{1}{3}$$

So the area of the given region is $\frac{2}{3}-\frac{1}{3}$, i.e., $\frac{1}{3}$. Checking, the area of the square bounded by the coordinate axes, $x=1$ and $y=1$ is 1, so that the area required is about $\frac{1}{3}$.

VOLUMES OF SOLIDS OF REVOLUTION

If a region bounded by the x-axis, the curve $y=f(x)$, and the lines $x=a$, $x=b$ is rotated completely about the x-axis, the volume of the solid so formed is $\pi\int_a^b y^2\,dx$ (Fig 23.3a); if the region bounded by the y-axis, $x=F(y)$, $y=c$ and $y=d$ is rotated completely about the y-axis, the volume of the solid so formed is $\pi\int_c^d x^2\,dy$.

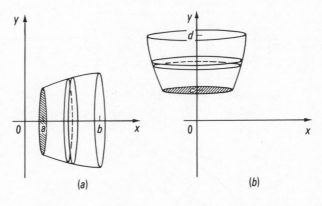

(a)   (b)

Fig 23.3

Example 23.9 The region bounded by $y=1+e^{-x}$, the x-axis, the y-axis and the line $x=1$ is rotated completely about the x-axis. Find the volume of the solid so formed (Fig 23.4).

$$V=\pi\int_0^1 y^2\,dx=\pi\int_0^1(1+e^{-x})^2\,dx$$

$$=\pi\int_0^1(1+2e^{-x}+e^{-2x})\,dx$$

$$=\left[x-2e^{-x}-\tfrac{1}{2}e^{-2x}\right]_0^1$$

$$\equiv 2.7\pi,\text{ about } 8.47$$

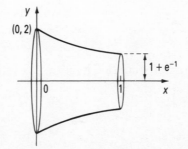

Fig 23.4

Checking, this solid is smaller than a circular cylinder radius 2, length 1, (volume $4\pi\simeq12$), greater than a cylinder radius $1+e^{-1}$*, and $\pi\times1.4^2$ is about 6, so the answer is reasonable.

*$1+e^{-1}\simeq1.4$

MEAN VALUE

If we have several numbers, e.g. 1,3,5,8, we can define their mean as $\frac{1}{4}(1+3+5+8)$, which gives a 'statistic' with which to describe this set of numbers. If we have a continuously-varying quantity described by $y=f(x)$, varying over an interval $a\leqslant x\leqslant b$, then we can define the mean value $\bar{y}$ of y over that interval as $\bar{y}=\dfrac{1}{b-a}\displaystyle\int_a^b y\,dx.$

Figure 23.5 illustrates the manner in which three different functions vary over the same interval. $y=x^2$ increases slowly at first; $y=x$ increases more rapidly initially, $y=\sin(\frac{1}{2}\pi x)$ increases even more rapidly initially. We should expect the corresponding values of y to have different means.

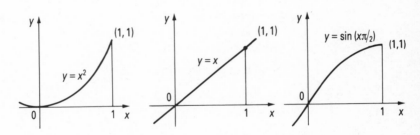

Fig 23.5

To find the mean value $\bar{y}$ of $y=x^2$ over the interval $0\leqslant x\leqslant1$,

$$\bar{y}=\frac{1}{1-0}\int_0^1 x^2\,dx$$

$$=\frac{1}{3}$$

the mean value of $y=x$ over the same interval is

$$\bar{y}=\int_0^1 x\,dx$$

$$=\tfrac{1}{2}$$

whereas of $y=\sin(\pi x/2)$, the mean value is

$$\int_0^1 \sin(\pi x/2)\,dx$$

$$=\left[-\frac{2}{\pi}\cos(\pi x/2)\right]_0^1$$

$$=\frac{2}{\pi}$$

greater than the others, as expected.

We can see from Fig 23.6 that the area of the region R_1 bounded by the x-axis, the ordinates $x=a$, $x=b$ and the curve $y=f(x)$ is equal to the area of the rectangle R_2, bounded by the x-axis, the ordinates $x=a$, $x=b$ and the line $y=\bar{y}$.

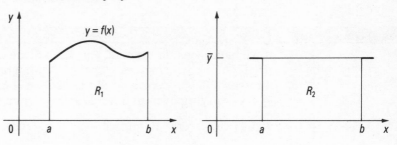

Fig 23.6

Find the integral with respect to x of

1   $\cos^2 x$

2   $\cos 3x \cos x$

3   $\sin 3x \cos 2x$

4   $\sin 5x \sin x$

5   $\sin^2 3x$

Evaluate the following integrals:

6   $\displaystyle\int_0^x \sin^2 \left(\tfrac{1}{2}x\right) dx$

7   $\displaystyle\int_0^{\pi/6} \sin 2x \cos x \, dx$

8   $\displaystyle\int_0^{\pi/2} \cos 2x \cos x \, dx$

9   $\displaystyle\int_0^{\pi/2} \sin 3x \sin x \, dx$

10   $\displaystyle\int_0^{\pi/2} \sin^4 x \, dx$

Integrate with respect to x:

11   $\dfrac{1}{x+2}$

12   $\dfrac{x}{x+2}$

13   $\dfrac{x^2}{x+2}$

14   $\dfrac{\sin x}{2-\cos x}$

15   $\dfrac{e^{2x}}{1+e^{2x}}$

Evaluate:

16 $\displaystyle\int_3^5 \frac{1}{(x+1)(x-2)}dx$

17 $\displaystyle\int_1^2 \frac{1}{x(x+3)}dx$

18 $\displaystyle\int_0^{\pi/2} \frac{\sin x}{1+\cos x}dx$

19 $\displaystyle\int_0^{\pi/6} \frac{\sin x+\cos x}{\cos x-\sin x}dx$

20 $\displaystyle\int_1^2 \frac{1}{x(x^2+1)}dx$

Find the following integrals:

21 $\displaystyle\int \frac{1}{\sqrt{(4-x^2)}}dx$

22 $\displaystyle\int \frac{1}{4+x^2}dx$

23 $\displaystyle\int \frac{1}{1+4x^2}dx$

24 $\displaystyle\int \frac{1}{\sqrt{(1-4x^2)}}dx$

25 $\displaystyle\int \frac{1}{25+9x^2}dx$

Evaluate:

26 $\displaystyle\int_0^{3/2} \frac{1}{\sqrt{(9-x^2)}}dx$

27 $\displaystyle\int_0^3 \frac{1}{9+x^2}dx$

28 $\displaystyle\int_0^{2/3} \frac{1}{\sqrt{(4-9x^2)}}dx$

29 $\displaystyle\int_{5/4}^{5/2} \frac{1}{\sqrt{(25-4x^2)}}dx$

30 $\displaystyle\int_0^{\infty} \frac{1}{25+4x^2}dx$

Using the substitutions given, find

31 $\displaystyle\int \frac{1}{x^2+4x+5}dx,\ z=x+2$

32 $\displaystyle\int \frac{1}{x^2+4x+3}dx,\ z=x+2$

33 $\displaystyle\int_0^3 x\sqrt{(1+x)}dx,\ z^2=1+x$

34 $\displaystyle\int_0^{\infty} \frac{1}{(1+x^2)^2}dx,\ \tan\theta=x$

35 $\int_0^1 \sqrt{\dfrac{x}{1-x}} dx$, $\sin^2 \theta = x$

Use integration by parts to find

36 $\int x \sin x \, dx$

37 $\int x \, e^x \, dx$

38 $\int x^2 \, e^x \, dx$

39 $\int x^2 \ln x \, dx$

40 $\int \arctan x \, dx$

41 Find the area of the region bounded by $y=x(2-x)$ and the x-axis. Find also the volume of the solid formed when this region is rotated completely about the x-axis.

42 Find the area of the finite region bounded by $y=3x-x^2$ and the straight line $y=x$. Find also the volume of the solid formed when this region is rotated completely about the x-axis.

43 The region bounded by the curve $y=\arcsin x$ and the y-axis for which $0 \leqslant y \leqslant \pi$ is rotated completely about the y-axis. Find (a) the area of this region, (b) the volume of the solid so formed.

44 The region bounded by $x^2+y^2=a^2$, the x-axis, the line $x=\tfrac{1}{2}a$ and that part of the x-axis for which $\tfrac{1}{2}a \leqslant x \leqslant a$ is rotated completely about the x-axis, to form a piece cut from a sphere. Find the volume of the solid so formed.

45 The region bounded by the y-axis, the x-axis, the curve $y=\ln x$ and the line $y=2$ is rotated completely about the y-axis. Find the volume of the solid so formed.

Find the mean value of each of the following, over the interval given.

46 $y=x^3$, $0 \leqslant x \leqslant 1$

47 $y=x^3$, $2 \leqslant x \leqslant 4$

48 $y=\sin x$, $0 \leqslant x \leqslant \pi/2$

49 $y=\sin x$, $0 \leqslant x \leqslant 2\pi$

50 $y=\sin^2 x$, $0 \leqslant x \leqslant 2\pi$

51 Integrate

(a) $\int \dfrac{1}{\sqrt{4-x}} \, dx$

(b) $\int \dfrac{1}{4-x} dx$

(c) $\int \dfrac{1}{\sqrt{(4-x^2)}} dx$

(d) $\int \dfrac{1}{4+x^2} \, dx.$

52 Use the substitution $z=e^x$ to find $\int \dfrac{e^x}{1+e^{2x}} \, dx.$

53 Using integration by parts, find $\int \arcsin x \, dx.$

54 Evaluate $\int_0^1 \dfrac{x^2}{(x+1)(x+3)} dx.$

55 The region R is bounded by the coordinate axes, the curve $y=\dfrac{1}{\sqrt{(1+x)}}$ and the straight line $x=1$. Find the area of R, and volume of the solid formed when R is rotated completely about (a) the x-axis, (b) the y-axis.

56 Show that $\displaystyle\int_0^{1/2} \dfrac{1}{1-x^2}dx=\tfrac{1}{2}\ln 3$.

57 The curves $y=1+x^3$ and $y=3x^2-3$ meet at $(-1,0)$ and at $(2,9)$. Sketch the two curves in the same diagram and show that the area of the region between the curves is 27/4.

58 Find the integral $\displaystyle\int\dfrac{1}{x(1-x^2)}dx$.

59 Find the mean value of $\sin^3 x$ over the interval $0\leqslant x\leqslant\pi/2$.

60 The displacement x of a particle at time t is given by $x=\sin t$. The mean value of its velocity with respect to time is $\dfrac{1}{b-a}\displaystyle\int_{t=a}^{t=b} v\,dt$; the mean value of its velocity with respect to displacement is $\dfrac{1}{d-c}\displaystyle\int_{x=c}^{x=d} v\,dx$. Calculate each of these mean values, over the interval $0\leqslant x\leqslant\pi/2$.

DIFFERENTIAL EQUATIONS

CONTENTS

Some differential equations can be solved by **integrating term by term** as they stand, e.g.

$$\frac{dy}{dx}=x+\cos 2x \Rightarrow y=\tfrac{1}{2}x^2+\tfrac{1}{2}\sin 2x+C$$

Some are such that the **variables are separable**, i.e. of the form $f(y)\dfrac{dy}{dx}=F(x)$, e.g.

$$\sin y\frac{dy}{dx} = \frac{1}{(1+x^2)} \Rightarrow \int \sin y\, dy = \int \frac{1}{1+x^2}dx$$

or are equations that can be reduced to this form.

When a **substitution** is given, do not forget to change $\dfrac{dy}{dx}$, e.g. if the substitution given is

$$y=vx,\ \frac{dy}{dx}=v+x\frac{dv}{dx}.$$

Second order differential equations will probably be of the form

$$\frac{d^2y}{dx^2}-k^2y=0,\text{ solution } y=Ae^{kx}+Be^{-kx}$$

or $$\frac{d^2y}{dx^2}+k^2y=0,\text{ solution } y=A\sin kx+B\cos kx$$

or can be reduced to this form.

DIFFERENTIAL EQUATIONS

It is most desirable to be able to see that a differential equation is one of a certain type, and in some cases it is possible to see at once which function satisfies the differential equation. For example, all different-ial equations of the type

$$\frac{dy}{dx}=f(x),\text{ e.g. } \frac{dy}{dx}=x+x^2+\sin x$$

can be integrated term by term.

Example 24.1 Solve $\dfrac{dy}{dx}=2x+x^3+\sin x$.

Integrating term by term,

$$y=x^2+\tfrac{1}{4}x^4-\cos x+C.$$

Since there is no information to determine the constant C, the final expression must be left in this form.

EXPONENTIAL FUNCTION

Given the equation $\dfrac{dy}{dx}=ky$, we know that the solution must be $y=Ce^{kx}$ for some constant C, for the exponential function is the only function f such that $\dfrac{d}{dx}(f)=kf$, and indeed it can be defined in that form. Similarly we shall see (p260) that we should quote at once the form of solutions of the equation

$$\dfrac{d^2y}{dx^2}=ky$$

and then find the value of the arbitrary constants that satisfy any initial (or other boundary) conditions given.

Example 24.2 Solve $\dfrac{dy}{dx}=-4y$, given that $y=3$ when $x=0$.

We know that $\dfrac{dy}{dx}=-4y\Rightarrow y=Ce^{-4x}$.

But when $x=0$, $y=3$ $\therefore$ $3=Ce^0$,

$$C=3,$$
So $\qquad y=3e^{-4x}$.

SEPARABLE VARIABLES　　Many differential equations can be rewritten in the form

$$f(y)\dfrac{dy}{dx}=F(x), \text{ e.g. } \sin y\dfrac{dy}{dx}=e^x$$

so that each side separately can be integrated, e.g. if

$$3y^2\dfrac{dy}{dx}=2x$$

then $\qquad y^3=x^2+C$

or if $\quad \sin y \dfrac{dy}{dx} = e^x$

then $\quad -\cos y = e^x + C$

Some may need rearranging before they are in a suitable form, e.g.

$$\frac{dy}{dx} = \frac{x}{(1+y^2)}$$

then $\quad (1+y^2)\dfrac{dy}{dx} = x$

whence

$$y + \tfrac{1}{3}y^3 = \tfrac{1}{2}x^2 + C,$$

whereas if

$$\frac{dy}{dx} = \frac{x}{1+xy}, \text{ then } (1+xy)\frac{dy}{dx} = x,$$

and we cannot solve this equation by this method, for we cannot find

$$\int xy \frac{dy}{dx} dx.$$

Example 24.3 Solve $\dfrac{1}{y}\dfrac{dy}{dx} + x = xy$, given that $y=2$ when $x=0$.

Rearranging, $\qquad\qquad \dfrac{1}{y}\dfrac{dy}{dx} = x(y-1)$

so that $\qquad\qquad \dfrac{1}{y(y-1)}\dfrac{dy}{dx} = x$

$$\int \frac{1}{y(y-1)} dy = \int x \, dx$$

$\therefore \qquad\qquad \displaystyle\int\left(-\frac{1}{y} + \frac{1}{(y-1)}\right) dy = \int x dx$

$\therefore \qquad\qquad \ln\left|\dfrac{y-1}{y}\right| = \tfrac{1}{2}x^2 + C$

But $y=2$ when $x=0$, so

$$\ln(\tfrac{1}{2}) = C, \text{ i.e. } C = -\ln 2$$

The solution is $\qquad \ln\dfrac{(y-1)}{y} = \tfrac{1}{2}x^2 - \ln 2$

i.e. $\qquad\qquad \dfrac{2(y-1)}{y} = e^{\frac{1}{2}x^2}$

EQUATIONS REQUIRING SUBSTITUTION

Some equations can be reduced to the form in which the variables are separable by means of a substitution, which will usually be given in examinations. For example,

if $\qquad \dfrac{dy}{dx}=x-y$

the variables x and y cannot be separated as the equation stands, but if we use the substitution

$$z=x-y$$

$$\frac{dz}{dx}=1-\frac{dy}{dx}, \text{ i.e. } \frac{dy}{dx}=1-\frac{dz}{dx}$$

the equation becomes

$$1-\frac{dz}{dx}=z$$

whence $\qquad \dfrac{dz}{dx}=1-z$

$$\int\frac{1}{1-z}dx=\int dx$$

i.e. $\qquad -\ln(1-z)=x+C$ $\hfill$ (1)

whence $\qquad 1-z=e^{-(x+C)}$

i.e. $\qquad 1-(x-y)=e^{-(x+C)}$

$$y=x-1-e^{-(x+C)}$$

Equation (1) illustrates that if the equation contains many logarithmic terms the constant of integration can often be more usefully taken as $\ln A$ (where here $-\ln A=C$), then

$$-\ln(1-z)=x-\ln A,$$

$$1-z=Ae^{-x}$$

and $\qquad y=x-1+Ae^{-x}$

Where it is necessary to use a substitution $z=f(x,y)$, care must always be taken to obtain $\dfrac{dz}{dx}$ correctly. The example below illustrates the method used when solving a homogeneous equation, one in which the terms are of the same degree in x and y, when the substitution $y=xz$ often reduces the equation to one in which the variables can be separated.

Example 24.4 Solve $xy\dfrac{dy}{dx}=x^2-y^2$.

Every term is of degree two (either x^2 or y^2 or x^1y^1) so we can try the substitution $y=xz$, where z is a function of x, so $\dfrac{dy}{dx}=z+x\dfrac{dz}{dx}$. The equation then becomes $x(xz)\left(z+x\dfrac{dz}{dx}\right)=x^2-x^2z^2$, i.e.

$$z\left(z+x\dfrac{dz}{dx}\right)=1-z^2$$

$$zx\dfrac{dz}{dx}=1-2z^2$$

$$\int\dfrac{-4z}{1-2z^2}dz=\int\dfrac{-4}{x}dx, \text{ multiplying by } -4$$

so that on the l.h.s. the numerator is the derivative of the denominator,

i.e. $\ln(1-2z^2)=-4\ln x+\ln A$

$$1-2z^2=Ax^{-4}$$

$$1-2\dfrac{y^2}{x^2}=\dfrac{A}{x^4}$$

$$x^4-2x^2y^2=A$$

SECOND ORDER DIFFERENTIAL EQUATIONS

We saw that the exponential function e^{kx} was the only function that satisfies the equation $\dfrac{d}{dx}(f)=kf$. We can see that $y=e^{kx}$ satisfies $\dfrac{d^2y}{dx^2}=k^2y$, but $y=e^{-kx}$ also satisfies the equation, and indeed so does $y=Ae^{kx}+Be^{-kx}$ for all values of A and B. It can be shown that this is the most general form for the solution of $\dfrac{d^2y}{dx^2}=k^2y$, and it is often wise to proceed straight to this solution, and find the constants A and B from the given data.

Example 24.5 Find the solution of $\dfrac{d^2y}{dx^2}=4y$, given that when $x=0$, $y=3$ and $\dfrac{dy}{dx}=2$.

We know that the general solution of $\dfrac{d^2y}{dx^2}=4y$ is

$$y=Ae^{2x}+Be^{-2x}$$

Since $y=3$ when $x=0$, $3=A+B$.

Since $\dfrac{dy}{dx}=2$ when $x=0$, $2=2A-2B$

Solving simultaneously, $A=2$ and $B=1$, so the solution required is
$$y=2e^{2x}+e^{-2x}$$

Example 24.6 Find the solution of $\dfrac{d^2y}{dx^2}=y$, given that $y=3$ when $x=0$, and that y becomes small as x becomes large and positive.

Again, we know that the general solution of $\dfrac{d^2y}{dx^2}=y$ is $y=Ae^x+Be^{-x}$, k^2 being equal to 1 in this case. Since $y=3$ when $x=0$, $3=A+B$, but the condition that $y\to0$ as $x\to+\infty$ requires that $A=0$, for otherwise $Ae^x\to\infty$ as $x\to+\infty$. Since $A=0$, $B=3$, and the required solution is
$$y=3e^{-x}$$

TRIGONOMETRIC FUNCTION

The differential equation $\dfrac{d^2y}{dx^2}=--=k^2y$ is clearly similar to $\dfrac{d^2y}{dx^2}=k^2y$, but here the solutions are of the form $y=A\sin kx+B\cos kx$, since $\sin kx$ and $\cos kx$ are the only functions that have the property $\dfrac{d^2y}{dx^2}=-k^2y$.

Again, write down the general form of the solution, and fit the initial or boundary conditions.

Example 24.7 Solve the differential equation $\dfrac{d^2y}{dx^2}=-25y$, given that when $x=0$, $y=0$ and $\dfrac{dy}{dx}=20$.

The general solution of the differential equation is

$$y=A\sin 5x+B\cos 5x$$

since $$k^2=25$$

Since $y=0$ when $x=0$,

$$0=B\cos 0$$

i.e. $$B=0$$

Since $\qquad$ when $x=0, \dfrac{dy}{dx}=20,$

$$20=5A \cos 0,$$

i.e. $\qquad A=4$

The solution is $y=4 \sin 5x$.

We soon come to recognize that if the initial conditions are such that $y=0$ when $x=0$, then $B=0$ and the solution is $y=A \sin kx$;

if when $x=0, \dfrac{dy}{dx}=0$, then $A=0$ and the solution is $y=B \cos kx$.

These are expected from our knowledge of the sine and cosine functions.

Example 24.8 Solve the differential equation $\dfrac{d^2y}{dx^2}+16y=0$, given that $y=1$ when $x=0$ and $y=5$ when $x=\pi/8$.

Here, $k^2=16$, so the solution is of the form

$$y=A \sin 4x+B \cos 4x$$

When $x=0$, $y=1$

$$1=B$$

when $x=\pi/8$, $y=5$

$$5=A \sin (\pi/2)+B \cos (\pi/2)$$

$$5=A$$

so the solution is

$$y=5 \sin 4x+\cos 4x$$

SECOND ORDER DIFFERENTIAL EQUATIONS WITH A CONSTANT

If the differential equation is of the form

$$\frac{d^2y}{dx^2}+ky=c$$

write it as $\qquad \dfrac{d^2y}{dx^2}+k\left(y-\dfrac{c}{k}\right)=0$

and substitute $z=y-c/k$.

Example 24.9 Solve $\dfrac{d^2y}{dx^2}+3y=12$.

Write the equation as

$$\frac{d^2y}{dx^2}+3(y-4)=0$$

Substituting $z=y-4$, $\dfrac{d^2z}{dx^2}=\dfrac{d^2y}{dx^2}$, the equation becomes

$$\frac{d^2z}{dx^2}+3z=0$$

The solution of this is

$$z=A\sin(x\sqrt{3})+B\cos(x\sqrt{3})$$

so the solution of the original equation is

$$y=A\sin(x\sqrt{3})+B\cos(x\sqrt{3})+4$$

It is to this final equation, of course, that we must fit the initial or boundary conditions.

1   Solve $\dfrac{dy}{dx}=3x^2+4x^3$, given that $y=1$ when $x=1$.

2   Solve $\dfrac{dy}{dx}=\sin x+e^x$, given that $y=0$ when $x=0$.

3   Solve $\dfrac{dy}{dx}=x^2+e^{-x}$, given that $y=0$ when $x=0$.

4   Solve $\dfrac{dy}{dx}=\cos 2x$, given that $y=0$ when $x=\pi/4$.

5   Solve $\dfrac{dy}{dx}=5y$, given that $y=2$ when $x=0$.

6   Solve $\dfrac{dy}{dx}=3y$, given that $\dfrac{dy}{dx}=6$ when $x=0$.

7   Solve $\dfrac{dy}{dx}+4y=0$, given that $y=5$ when $x=0$.

8   Solve $\dfrac{dy}{dx}+y=0$, given that $y=1$ when $x=2$.

9   Solve $x\dfrac{dy}{dx}=y^2$, given that $y=1$ when $x=1$.

10   Solve $x\dfrac{dy}{dx}=\cot y$, given that $y=\dfrac{\pi}{2}$ when $x=1$.

11   Solve $\cot x\dfrac{dy}{dx}=1+y^2$, given that $y=1$ when $x=0$.

12   Solve $\cot x\dfrac{dy}{dx}=1-y^2$, given that $y=0$ when $x=\dfrac{\pi}{4}$.

13   Solve $x\dfrac{dy}{dx}=(1+2x^2)y^2$, given that $y=1$ when $x=1$.

14   Solve $\dfrac{dy}{dx}=x+y$, using the substitution $z=x+y$.

15 Solve $\dfrac{dy}{dx}=(x+y)^2$, given that $y=1$ when $x=0$. [Use the substitution $z=x+y$.]

16 Solve $xy\dfrac{dy}{dx}=x^2+y^2$, given that $y=1$ when $x=1$.

17 Solve $xy\dfrac{dy}{dx}=x^2-y^2$, given that $y=0$ when $x=1$.

Solve the following differential equations.

18 $\dfrac{d^2y}{dx^2}-4y=0$, given $y=2$ and $\dfrac{dy}{dx}=0$ when $x=0$.

19 $\dfrac{d^2y}{dx^2}+4y=0$, given $y=2$ and $\dfrac{dy}{dx}=0$ when $x=0$.

20 $\dfrac{d^2y}{dx^2}-9y=0$, given $y=0$ and $\dfrac{dy}{dx}=6$ when $x=0$.

21 $\dfrac{d^2y}{dx^2}+9y=0$, given that $y=0$ and $\dfrac{dy}{dx}=6$ when $x=0$.

22 $\dfrac{d^2y}{dx^2}+y=0$, given that $y=3$ and $\dfrac{dy}{dx}=1$ when $x=0$.

23 $\dfrac{d^2y}{dx^2}-y=0$, given that $y=3$ and $\dfrac{dy}{dx}=1$ when $x=0$.

24 $4\dfrac{d^2y}{dx^2}+y=0$, given that $y=3$, $\dfrac{dy}{dx}=0$ when $x=\pi$.

25 $4\dfrac{d^2y}{dx^2}-y=0$, given that $y=3$, $\dfrac{dy}{dx}=0$ when $x=1$.

26 $25\dfrac{d^2y}{dx^2}+4y=0$, given that $y=5$, $\dfrac{dy}{dx}=1$ when $x=0$

27 $25\dfrac{d^2y}{dx^2}-4y=0$, given that $y=5$, $\dfrac{dy}{dx}=2$ when $x=0$.

28 $\dfrac{d^2y}{dx^2}+4y=8$, $y=3$ and $\dfrac{dy}{dx}=0$ when $x=0$.

29 $\dfrac{d^2y}{dx^2}-4y=8$, $y=4$ and $\dfrac{dy}{dx}=0$ when $x=0$.

30 Solve $x\dfrac{dy}{dx}=(1+x)y$, given that $y=e$ when $x=1$.

31 Solve $(1+2x^2)\dfrac{dy}{dx}=xy$, given that $y=1$ when $x=0$.

32 Solve $\dfrac{dy}{dx}+xy^3=y^3$, given that $y=1$ when $x=1$.

33 Solve $\dfrac{dy}{dx}=\tfrac{1}{2}xy$, given that $y=1$ when $x=2$.

34 Solve $\dfrac{dy}{dx}+x^2+4x^3=0$, given that $y=5$ when $x=1$.

35 Solve $\dfrac{d^2y}{dx^2}+4\dfrac{dy}{dx}=0$.

36 Solve $\dfrac{d^2y}{dx^2}+4\dfrac{dy}{dx}=8$.

37 Solve $x\dfrac{dy}{dx}=y(y-2)$.

38 Solve $\dfrac{dy}{dx}=xy(y-2)$.

39 Solve $\cos x\dfrac{dy}{dx}-y\sin x=y^2\sin x$.

NUMERICAL METHODS

CONTENTS

NOTES

APPROXIMATE INTEGRATION

If $y_0, y_1, y_2, \ldots y_n$ are successive ordinates dividing a region R into n strips of equal width h, the trapezium rule says that the area A of R is approximately

$$\tfrac{1}{2}h\{y_0+2(y_1+y_2+y_3\ldots)+y_n\}$$

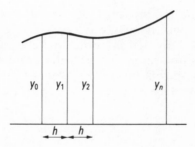

Fig 25.1

Simpson's rule says that, **when n is even,**

$$A \approx \tfrac{1}{3}h\{y_0+4y_1+2y_2+4y_3+2y_4\ldots+4y_{n-1}+y_n\}.$$

MACLAURIN'S THEOREM

If all derivatives of $f(x)$ exist when $x=0$,

$$f(x)=f(0)+f'(0)+\frac{1}{2!}x^2 f''(0)+\frac{1}{3!}x^3 f'''(0)+ \ldots$$

TAYLOR'S THEOREM

If all derivatives of $f(x)$ exist when $x=a$,

$$f(a+x)=f(a)+xf'(a)\frac{1}{2!}+x^2 f''(a)+\frac{1}{3!}x^3 f'''(a)+ \ldots$$

NEWTON–RAPHSON FORMULA

If x_r is a good approximation to a root of $f(x)=0$, a better approximation is x_{r+1}, where

$$x_{r+1}=x_r-\frac{f(x_r)}{f'(x_r)}$$

OTHER ITERATIONS

If the equation $f(x)=0$ is to be solved by an iteration

$$x_{r+1}=F(x_r)$$

the iteration will converge to the required root if $|F'(x_r)|<1$.

APPROXIMATE INTEGRATION

Considering the region bounded by the curve $y=f(x)$, the straight lines $x=a$, $x=b$ and the x-axis, we see that the area of each strip into which it can be divided is approximately $\frac{1}{2}h(y_k+y_{k+1})$, h being the width of the strip and y_k and y_{k+1} the ordinates at each side of the strip. Continuing this procedure over the whole of the region, the area is approximately

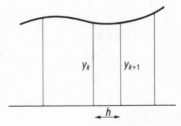

Fig 25.2

$$\tfrac{1}{2}h(y_0+2y_1+2y_2+2y_3+ \ldots +2y_{n-1}+y_n)$$

if we have divided the region into n trapezia. This is called the **trapezium rule**.

SIMPSON'S RULE

The equation of a parabola with axis parallel to the y-axis is $y=ax^2+bx+c$, for some a, b and c. Considering a region bounded by such a parabola, straight lines $x=-h$, $x=h$ and the x-axis, the area A is *exactly* equal to $\int_{-h}^{h}(ax^2+bx+c)\mathrm{d}x$, i.e.

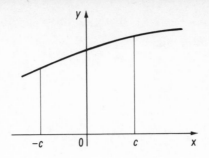

Fig 25.3

$$A=\left[\tfrac{1}{3}ax^3+\tfrac{1}{2}bx^2+cx\right]_{-h}^{h}$$

$$=\tfrac{2}{3}ah^3+2ch$$

$$=\tfrac{1}{3}h(2ah^2+6c)$$

Denoting the ordinates by y_0, y_1, y_2,

we see
$$y_0=ah^2-bh+c,$$
$$y_1=c \text{ and }$$
$$y_2=ah^2+bh+c,$$

so that
$$y_0+4y_1+y_2=2ah^2+6c$$
i.e.
$$A=\tfrac{1}{3}h(y_0+4y_1+y_2)$$

and the area of this region is exactly equal to $\tfrac{1}{3}h(y_0+4y_1+y_2)$.

It can be shown that this is true for all parabolae with axes parallel to the y-axis, so that if we divide a region into parts, each width $2h$, each approximately bounded on one side by a parabola, then the area of the region is *approximately*

$$\tfrac{1}{3}h(y_0+4y_1+2y_2+4y_3+ \ldots +2y_{2n-2}+4y_{2n-1}+y_{2n})$$

Notice that we have to have an even number of ordinates, since the width of the 'parabolic strips' is $2h$.

Many calculators are now programmable so that they will carry out the required calculations for the trapezium rule or for Simpson's rule, and even those that cannot be so programmed will carry out much of the work done in detail in this example. In an examination, though, candidates are always required to show enough working for the examiner to be able to see that he or she has understood the method used.

Example 25.1 Find $\int_0^1 e^{-x^2}dx$, using (*a*) the trapezium rule, (*b*) Simpson's rule, with 10 intervals.

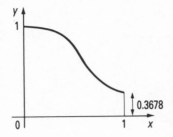

Fig 25.4

Forming a table of values, we have

x	0	0.1	0.2	0.3	0.4	0.5	0.6
e^{-x^2}	1	0.9900	0.9608	0.9139	0.8521	0.7788	0.6977
x	0.7	0.8	0.9	1.0			
e^{-x^2}	0.6126	0.5273	0.4449	0.3678			

(*a*)   Using the trapezium rule,

$A \approx \frac{1}{2}(0.1)[1+2(0.9900+0.9608+0.9139+0.8521+0.7788$
$+ 0.6977+0.6126+0.5273+0.4449)+0.3678]$
≈ 0.746

(*b*)   Using Simpson's rule,

$A \approx \frac{1}{3}(0.1)[1+4(0.9900+0.9139+0.7788+0.6126+0.4449)$
$+2(0.9608+0.8521+0.6977+0.5273)+0.3678]$
≈ 0.747

Checking, we expect the area to be slightly greater than that of a trapezium, width 1, parallel sides 1 and 0.3678, i.e. greater than 0.6839.

These approximate methods are useful when we have a non-integrable function, as here, or in cases where we only know the ordinates, and not the function that describes the relation (see question 3 below).

MACLAURIN'S THEOREM

We should be familiar with certain approximations that are convenient for small values of x, e.g. $\sin x \approx x$, $\cos x \approx 1-\frac{1}{2}x^2$. Maclaurin's theorem enables us to find polynomials that are good approximations to other functions when x is small.

Maclaurin's theorem states that if $f(x)$ is a function of x for which all derivatives exist and have finite values when $x=0$,

$$f(x) \equiv f(0)+xf'(0)+\frac{1}{2!}x^2f''(0)+\frac{1}{3!}x^3f'''(0)\ldots$$

To prove this, suppose that

$$f(x) \equiv a_0 + a_1 x + \frac{1}{2!}a_2 x^2 + \frac{1}{3!}a_3 x^3 + \ldots$$

(The working is slightly easier if we introduce the factorials in the denominators.) Then when $x=0$, $f(0)=a_0$.

Differentiating $\qquad f'(x) = a_1 + a_2 x + \frac{a_3}{2!}x^2 + \frac{a_4}{3!}x^3 \ldots$

When $\qquad\qquad\qquad x=0, f'(0)=a$

Differentiating again, $\quad f''(x) = a_2 + a_3 x + \frac{1}{2!}a_4 x^2 + \frac{1}{3!}a_5 x^3 + \ldots$

Again putting $x=0$, $\quad f''(0)=a_2$. Proceeding in this way we obtain $a_0=f(0)$, $a_1=f'(0)$, $a_2=f''(0)$, $a_3=f''(0)$, $\ldots$

giving

$$f(x) = f(0) + xf'(0) + \frac{1}{2!}xf''(0) + \frac{1}{3!}x^3 f'''(0) + \ldots$$

The need that the derivatives should exist when $x=0$ is illustrated when trying to find an expansion for $\ln x$. For if $f(x)=\ln x$, $f'(x)=\frac{1}{x}$, $f''(x)=-\frac{1}{x^2}$ ... and $f'(0)$, $f''(0)$ and all later derivatives are undefined when $x=0$.

Example 25.2 Use Maclaurin's theorem to find the expansion, up to the term in x^2, of $\ln(1+\sin x)$.

Since $f(x)=\ln(1+\sin x)$, $f(0)=\ln 1=0$.

$$f'(x)=\frac{\cos x}{1+\sin x}, f'(0)=1$$

$$f''(x)=\frac{(1+\sin x)(-\sin x)-\cos x(\cos x)}{(1+\sin x)^2}$$

$$\dot{} = \frac{-1-\sin x}{(1+\sin x^2)} = \frac{-1}{1+\sin x}$$

so $\qquad f''(0)=-1$

$$\ln (1+\sin) \approx x - \tfrac{1}{2}x^2$$

Check. When $x=0.1$ (radians, of course), $\ln (1+\sin x)=0.09516$; $x-\tfrac{1}{2}x^2=0.095$.

TAYLOR'S THEOREM

Taylor's theorem is a more general form of Maclaurin's theorem, stating that if all derivatives of $f(x)$ exist when $x=a$,

$$f(x+a)\equiv f(a)+xf'(a)+\frac{1}{2!}x^2f''(a)+\frac{1}{3!}x^3f'''(a)+ \ldots$$

It is easily seen that Maclaurin's theorem is the special case of Taylor's theorem when $a=0$. The proof is similar to that of Maclaurin's theorem.

Example 25.3 Use Taylor's theorem to find the expansion, up to and including the term in x^3, of $\sin(x+\pi/4)$ in ascending powers of x.

Since $f(x)=\sin x$, $f(\pi/4)=--\frac{1}{\sqrt{2}}$; since $f'(x)=\cos x$, $f'(\pi/4)=\frac{1}{\sqrt{2}}$; since $f''(x)=-\sin x$, $f''(\pi/4)=-\frac{1}{\sqrt{2}}$; since $f'''(x)=-\cos x$, $f'''(\pi/4)=-\frac{1}{\sqrt{2}}$.

$$\therefore \qquad \sin(x+\pi/4)\approx\frac{1}{\sqrt{2}}\left(1+x-\frac{1}{2!}x^2-\frac{1}{3!}x^3\right)$$

Again, check the answer by putting $x=0.1$: $\sin(0.1+\pi/4)\approx0.77417$; the expansion ≈0.77416.

NEWTON–RAPHSON METHOD FOR FINDING SUCCESSIVE APPROXIMATIONS TO A ROOT OF AN EQUATION

If $x=a$ is a good approximation to a root of an equation $f(x)=0$, the exact value of that root will be $a+\varepsilon$, where ε is small, so that $f(a+\varepsilon)=0$. By Taylor's theorem,

$$f(a+\varepsilon)=f(a)+\varepsilon f'(a)+ \ldots$$
$$\approx f(a)+\varepsilon f'(a), \text{ since } \varepsilon \text{ is small.}$$

Now since $f(a+\varepsilon)=0$, $\varepsilon=\frac{-f(a)}{f'(a)}$, so that we can deduce that, if $x=a$ is a good approximation to a root of the equation $f(x)=0$,

$$x=a-\frac{f(a)}{f'(a)} \text{ is a better approximation.}$$

This is illustrated by Fig 25.5, which assumes that $f(a)$ and $f'(a)$ are both positive. (The figure can be adapted for the cases when either or both are negative.) If R is the point where $x=a$, $PR=f(a)$, the gradient at P is $f'(a)$, so that $QR=PR/\tan\psi=f(a)/f'(a)$ and the x-coordinate of Q is $a-f(a)/f'(a)$. It can be seen that Q is nearer to the point at which the curve crosses the x-axis than R, so gives a closer approximation to the root of $f(x)=0$.

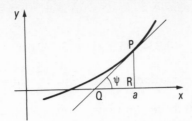

Fig 25.5

Since we can use this method to find successive approximations, we generally adopt the suffix notation, writing

if x_r is a good approximation to a root of the equation $f(x)=0$,

$$x_{r+1}=x_r-\frac{f(x_r)}{f'(x_r)}$$ is a better approximation.

To find a first approximation, we find two values of x for which $f(x)$ differ in sign; we may use linear interpolation to find a fairly good approximation with which to start.

N.B. The method fails if $f'(a)$ is near to zero, or if $f''(a)$ is very large.

Example 25.4 Find the root of $x^3-3x-6=0$, correct to 3sf.

Tabulating the values of $f(x)$, to find a first approximation,

$$f(0)=-6, f(1)=-8, f(2)=-4, f(3)=12,$$

so there is a root between $x=2$ and $x=3$. Using Fig 25.6, we see that $x=2.25$ is likely to be a good approximation to the root. (Since the triangles are similar with sides in the ratio 4:12, AB:BC=4:12, giving AB=0.25.)

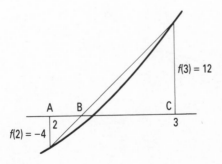

Fig 25.6

Since $f(x)=x^3-3x-6, f'(x)=3x^2-3,$

$$f(2.25)=-1.359\,375, f'(x)=12.187\,5,$$

a better approximation is $2.25-\dfrac{-1.359\,375}{12.187\,5}=2.361\,538.$

Repeating the procedure there is no need to work with many

significant figures, because it is unlikely that the later ones will be correct.

$$f(2.361\,5)=0.084\,835, f'(x)=13.730\,046,$$

so a better approximation is $2.361\,5-\dfrac{0.084\,835}{13.730\,046}=2.355\,32.$

Since each of these approximations is equal to 2.36, to 3sf, we know that the root has the required degree of accuracy.

The wording of the question indicated that there was only one real root of this cubic. This can be confirmed by noticing that the turning values occur where $3x^2-3=0$, i.e. $x=-1$ or 1, and that $f(-1)$ and $f(1)$ are both negative.

OTHER ITERATIVE METHODS

We can probably recall drawing graphs of say $y=1/x$ and $y=(x-1)(x+2)$ to find a solution of $x(x-1)(x+2)=1$, since the equation can be rearranged as $(x-1)(x+2)=1/x$. Many equations can be rearranged in the form $x=F(x)$, e.g. $x^3-x-2=0$ as $x=(x+2)^{1/3}$ and a method can be devised for finding successive approximations to a root of the equation.

Figure 25.7 shows the graphs of $y=x$ and $y=F(x)$, intersecting at a point X, whose x-coordinate is the root of the equation $x=F(x)$. If x_r is a good approximation to the root we see that by following the lines marked $\longrightarrow$ we can proceed to the point X. The y-coordinate of P is $f(x_r)$ and since Q lies on the line $y=x$, the x-coordinate of Q is $f(x_r)$, which is a better approximation to the root of the equation than x_r. Thus we can say that, under certain circumstances,

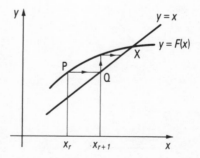

Fig 25.7

if x_r is a good approximation of a root of the equation $x=F(x)$, then $F(x_r)$ is a better approximation,

which can be written as an algorithm

$$x_{r+1}=F(x_r)$$

By drawing various curves $y=F(x)$ we can see that in many cases lines like $\longrightarrow$ will not approach the point of intersection (Fig 25.8), and it can be shown that the condition that an iteration of this form converges to a root is that $|F'(x_0)|<1$; the smaller $|F'(x_0)|$, the more rapid the convergence.

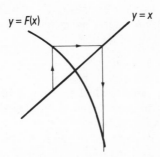

$y = F(x)$ $y = x$

Fig 25.8

Example 25.5 The equation $x^3-2x-3=0$ has a root near to $x=2$. The equation can be rearranged as

$$x=\tfrac{1}{2}(x^3-3) \text{ or as } x=\sqrt[3]{(2x+3)}.$$

By carrying out two iterations in each case show that $x_{r+1}=\tfrac{1}{2}(x_r^3-3)$ does not approach the root, but that $x_{r+1}=\sqrt[3]{(2x_r+3)}$ does.

Using $x_{r+1}=\tfrac{1}{2}(x_r^3-3)$,
if $x_0=2$, $x_1=\tfrac{1}{2}(2^3-3)=2.5$
                  $x_2=\tfrac{1}{2}(2.5^3-3)=6.3215$

which is not approaching the root.

Using $x_{r+1}=\sqrt[3]{(2x_r+3)}$,
if $x_0=2$, $x_1=\sqrt[3]{7}\approx1.913$
                  $x_2\approx1.897$
                  $x_3\approx1.894$

and clearly $x=1.89$ will be the value of the root to 3 sf.
    Notice that if $F(x)=\tfrac{1}{2}(x^3-3)$, $F'(x)=\tfrac{3}{2}x^2$ and $F'(2)=6$, which is greater than 1, whereas if $F(x)=(2x+3)^{1/3}$, $F'(x)=\tfrac{2}{3}(2x+3)^{-2/3}$, and $F'(2)\approx0.18$, which is much less than 1.

QUESTIONS

1   Using (a) the trapezium rule, (b) Simpson's rule, in each case with 10 intervals, find approximations to

$$\int_0^1 \frac{1}{1+x}\,dx$$

and check your answers by direct integration.

2 Using (a) the trapezium rule, (b) Simpson's rule, in each case with 10 intervals, find approximations to

$$\int_0^{\pi/2} \sqrt{\sin x}\, dx.$$

3 Given that $y=f(x)$, and the values of y at equal intervals of x as below, find (a) using the trapezium rule, (b) using Simpson's rule, the approximate area of the region bounded by the x-axis, the y-axis, the line $x=10$ and the curve $y=f(x)$.

x	0	1	2	3	4	5	6	7	8	9	10
y	0	3	7	9	5	6	7	7	8	2	1

4 Use Maclaurin's theorem to find the expansions of $\sin x$ and $\cos x$, giving the first two non-zero terms in each case.
Deduce $(x+\sin x)\cos x = 2x + \frac{5}{6}x^3$,
if powers of x higher than x^3 are neglected.

5 Use Maclaurin's theorem to find the expansion of $\ln(1+x)$ up to and including the term in x^4, and deduce that

$$(1+x)^2 \ln(1+x) = x + \tfrac{3}{2}x^2 + \tfrac{1}{3}x^3 - \tfrac{1}{12}x^4$$

if powers of x higher than x^4 are neglected.

6 Find the expansion, up to and including the term in x^4, of $e^{\sin x}$.

7 Find the first three non-zero terms in the expansion of $e^x \sin x$.

8 By finding the expansions of $(1-x^2)^{\frac{1}{2}}$ and arc sin x, show that $(1-x^2)^{1/2}$ arc sin $x \approx x - \tfrac{1}{3}x^3$.

9 Show that there is a root between 0 and 1 of

$$x^3 - 5x + 2 = 0$$

and find this root correct to 3dp.

10 Show that there is a root between 0 and 1 of

$$x^3 - 6x = -1$$

and find this root correct to 3dp.

11 Find the root between $x=2$ and $x=3$ of the equation $x^3 - 3x - 4 = 0$, starting with one linear interpolation and then using two iterations of Newton's formula.

12 Find an approximation to the positive root of $x^4 + x^2 - 19 = 0$, correct to 4 sf.

13 Find, correct to 3 sf, the root near to $x=1.2$ of $\sin x + 3\cos x = 2$.

14 Use the iteration

$$x_{r+1} = 5 - 2/x_r$$

to find the root near to 4 of the equation

$$x^2 - 5x + 2 = 0,$$

giving this root correct to 3dp.

15   Use the iteration

$$x_{r+1} = (1/6)(x^3_r + 2)$$

to find the solution of

$$x^3 - 6x + 2 = 0$$

near to 0.5, giving the root correct to 4dp. Why will this iteration not give the root near to 2? Find an iteration to obtain this root correct to 2dp.

16   The equation $x^3 - 10x + 1 = 0$ has roots near to -3, near to 0 and near to $+3$. Find algorithms to obtain these roots, and calculate approximations to these roots, correct to 2dp.

17   The equation $\tan x = 2x$ has a root near to $x = 1$. Show that $x_{r+1} = \frac{1}{2} \tan x_r$ will not obtain that root, but that $x_{r+1} = \arctan 2x_r$ will. Use this iteration to obtain that root, correct to 2dp. Compare this method with Newton's method by also solving the equation using Newton's method.

18   The equation $xe^x = 6$ has a root close to 1.5. Show that $x_{r+1} = 6e^{-x_r}$ is not a suitable algorithm, and find one that can be used to obtain this root. Use the algorithm to find the root correct to 3sf.

DISPLACEMENT, VELOCITY, ACCELERATION

CONTENTS

NOTES

Displacement describes the position of a particle relative to an origin O; the **distance** from O is the magnitude of the displacement.
Velocity is rate of change of displacement with respect to time.
Acceleration is rate of change of velocity with respect to time.

CONSTANT ACCELERATION FORMULAE

$v = u + at$
$s = ut + \frac{1}{2}at^2$
$s = \frac{1}{2}(u+v)t$
$v^2 = u^2 + 2as$

VARIABLE ACCELERATION

If the force (or acceleration) is a function of **time** t, use $\frac{dv}{dt} = f(t)$, $v = \frac{ds}{dt} = \int f(t)dt$ and integrate term by term.

If the force is a function of the **velocity**, use $\frac{dv}{dt} = F(v)$ and separate the variables to find v in terms of t; to find x in terms of t, integrate again. To find v in terms of x, use $v\frac{dv}{dx} = F(v)$ and separate the variables.

If the force is a function of the **displacement** x, use $v\frac{dv}{dx} = G(x)$ and separate the variables.

DISPLACEMENT AND DISTANCE

The **displacement** of a particle describes its position relative to a fixed origin O. In two or more dimensions, we use **r** to describe the displacement. In one dimension, if the x-axis is taken along the line of motion of the particle, x is the displacement of the particle O, $|x|$ is the distance of the particle from O.

VELOCITY AND SPEED

The velocity $\mathbf{v}$ of a particle is the rate of change of displacement; i.e., $\mathbf{v}=\dfrac{d}{dt}(\mathbf{r})$; the speed v of a particle is $|\mathbf{v}|$.

ACCELERATION

The acceleration $\mathbf{a}$ of a particle is the rate of change of velocity, i.e.

$\mathbf{a}=\dfrac{d}{dt}(\mathbf{v})$.

CONSTANT ACCELERATION FORMULAE

If a particle is moving in a straight line with constant acceleration a, $\dfrac{dv}{dt}=a \Rightarrow v=at+C$. If the velocity is u when $t=0$, $u=C$ and the equation is

$$v=u+at \tag{1}$$

Writing s to denote the displacement of the particle,

$$\dfrac{ds}{dt}=u+at \Rightarrow s=ut+\tfrac{1}{2}at^2+C$$

If we measure $s=0$ when $t=0$, $C=0$ and the equation becomes

$$s=ut+\tfrac{1}{2}at^2 \tag{2}$$

Equation (2) can be rewritten
$$s=\tfrac{1}{2}(2u+at)t$$
$$=\tfrac{1}{2}[u+(u+at)]t$$

Since $v=u+at$,

$$s=\tfrac{1}{2}(u+v)t \tag{3}$$

From (1), $t=\dfrac{v-u}{a}$, so that, from (3),

$$s=\tfrac{1}{2}(u+v)\dfrac{(v-u)}{a}$$

i.e. $2as=v^2-u^2$,
$$v^2=u^2+2as \tag{4}$$

N.B. *These equations are only true if the particle is moving in a straight line with constant acceleration.*
When solving problems using these equations, always tabulate the data to see which quantities we know; we need to know three quantities, then we can always find the other two.

Example 26.1 A particle is moving in a straight line with constant acceleration. If the initial velocity is 7m s^{-1} and the velocity after 10 s is 12m s^{-1}, find the acceleration, and the displacement of the particle after 100 s.

Here $u=7$ $v=12$ $s=?$ $a=?$ $t=10$

To find a, use $v=u+at$, i.e. $12=7+10a$, $a=0.5$, the acceleration is 0.5m s^{-2}.

To find s when $t=100$,

we have $u=7$ $v=?$ $s=?$ $a=0.5$ $t=100$

so use $s=ut+\frac{1}{2}at^2$, i.e. $s=700+\frac{1}{2}(0.5)100^2=3200$

the displacement is 3200m.

GRAPHICAL ILLUSTRATIONS

Since $v=\dfrac{ds}{dt}$, the gradient at a point on a displacement–time curve gives the velocity at that time t.

Since $a=\dfrac{dv}{dt}$ and $s=\displaystyle\int\frac{ds}{dt}dt=\int vdt$ a velocity–time curve is often much more useful, for the gradient at any point (time t_1) gives the velocity at time t_1; the area between $t=t_1$ and $t=t_2$ gives the displacement s in the interval in which t has changed from t_1 to t_2.

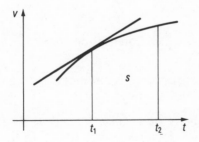

Fig 26.1

Example 26.2 A car travels 4km in 2 minutes, from rest to rest. It accelerates with constant acceleration to its maximum velocity of 40m s^{-1}, which it holds until it slows down, with constant deceleration. The time accelerating is three times that decelerating. Find for how long the car is accelerating.

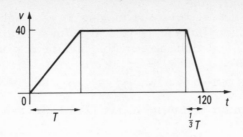

Fig 26.2

If the time spent accelerating is T seconds, the time decelerating is $\frac{1}{3}T$ seconds and the time at constant velocity is $(120-\frac{4}{3}T)$s. Considering the area under the velocity–time graph,

$$\tfrac{1}{2}\times40T+40(120-\tfrac{4}{3}T)+\tfrac{1}{2}\times40\times\tfrac{1}{3}T=4000$$
$$120-\tfrac{2}{3}T=100$$
$$T=30$$

The time taken accelerating is 30 s.

VARIABLE ACCELERATION

If a particle of constant mass is acted on by a force that varies, then since the relation between force and acceleration is described by Newton's second Law, i.e.

force=mass×acceleration,

the acceleration varies, and the equations (1) to (4) given on p284 are not true. We have to return to our definition of acceleration.

If **the force is a function of time**, use $a=\dfrac{dv}{dt}$ to obtain v as a function of t, then $v=\dfrac{ds}{dt}$ to obtain s in terms of t.

Example 26.3 A body mass 10kg is acted on by a force $5(1-e)^{-t/10}$)N. If the body initially has velocity of 2m s^{-1} in the direction of the force, find the velocity after 5 s, and the displacement of the body at the end of the 5 s.

Since $F=ma$

$$5(1-e^{-t/10})=10\frac{dv}{dt}$$

i.e. $$2\frac{dv}{dt}=1-e^{-t/10}$$
$$v=\tfrac{1}{2}(t+10e^{-t/10})+C$$

But $v=2$ when $t=0$, so $2=\frac{1}{2}(10)+C \Rightarrow C=-3$

$$v=\tfrac{1}{2}(t+10e^{-t/10}-3),$$

and the velocity after 5 s is $\frac{1}{2}(2+10e^{-\frac{1}{2}})$m s^{-1}, about 4.03m s^{-1}.

If the force is a function of the velocity, we still use $a=\dfrac{dv}{dt}$ but probably have to solve a differential equation with separable variables.

Example 26.4 A body mass 2kg is acted on by a force $(8-v)$ N, where vm s^{-1} is the velocity of the body. If the body is initially at rest, find its velocity and the displacement after (a) 1 s, (b) 10 s.

Using $F=ma$ again,

$$8-v=2\frac{dv}{dt},$$

$$\frac{1}{8-v}\frac{dv}{dt}=\tfrac{1}{2}$$

$$-\ln(8-v)+C=\tfrac{1}{2}t$$

But when $t=0$, $v=0$ so $C=\ln 8$, and the equation of motion is

$$\ln 8-\ln(8-v)=\tfrac{1}{2}t$$

$$\ln\left(\frac{8}{8-v}\right)=\tfrac{1}{2}t$$

$$\frac{8}{8-v}=e^{\frac{1}{2}t}$$

$$8=(8-v)e^{\frac{1}{2}t}$$

$$v=8(1-e^{-\frac{1}{2}t})$$

After 1 s, the velocity is $8(1-e^{-\frac{1}{2}})$ m s^{-1}, about 3.1m s^{-1}; after 10 s, the velocity is $8(1-e^{-5})$m s^{-1}, very close to 8m s^{-1}, and we can see that the velocity approaches 8m s^{-1} as t becomes large. This is called the **limiting velocity**.

To find s, use $v=\dfrac{ds}{dt}$

So $\qquad \dfrac{ds}{dt}=8(1-e^{-\frac{1}{2}t})$

$$s=8t+16e^{-\frac{1}{2}t}+C$$

When $t=0$, $s=0$ since we almost always measure s from the point where $t=0$, so $0=16+C$, $C=-16$, and the displacement s is given by

$$s=8t+16e^{-\frac{1}{2}t}+C$$

Now when $t=1$, $s=8-16(1-e^{-\frac{1}{2}})$, about 1.7, so the displacement after 1 s is about 1.7m; when $t=10$, $s=80-16(1-e^{-5})$, about 64, and the displacement is about 64m. Notice that of course this is less than if the body was travelling all the time at its limiting velocity.

FORCE AS A FUNCTION OF DISPLACEMENT

If the force acting on a body is a function of the displacement x, the acceleration and the velocity will also be functions of x. Then $\dfrac{dv}{dt} = \dfrac{dv}{dx} \cdot \dfrac{dx}{dt} = v\dfrac{dv}{dx}$, and this is the form that we have to use for the acceleration.

Example 26.5 The acceleration of a particle moving in a straight line is $(-4x)$m s^{-2}, where the displacement is xm and the force is directed towards the origin. If initially the particle is at rest a distance 10m from O, find the velocity with which it passes through O.

Taking the acceleration as $v\dfrac{dv}{dx}$,

$$v\dfrac{dv}{dx} = -4x,$$
$$\tfrac{1}{2}v^2 = -2x^2 + C$$

But $v=0$ when $x=10$, so $C=200$. The equation is

$$v^2 = 4(100 - x^2)$$

and $v=20$ when $x=0$, so the particle passes through O with velocity 20m s^{-1}.

If we wish to find a relation between x and t, write $v = \dfrac{dx}{dt}$, then $\dfrac{dx}{dt} = -2\sqrt{(100 - x^2)}$, taking the negative square root since the particle is moving towards O.

Then $\displaystyle\int \dfrac{-1}{\sqrt{(100 - x^2)}}\, dx = 2dt$
$$\text{arc cos}\left(\dfrac{x}{10}\right) = 2t + C.$$

But when $t=0$, $x=10$, so $C=0$, and the relation is

$$x = 10 \cos 2t.$$

Notice that we use $\displaystyle\int \dfrac{-1}{\sqrt{(100 - x^2)}}\, dx = \text{arc cos } (x/10)$. Taking the principal value, the graph shows that $\dfrac{d}{dx}$ (arc cos x) is negative, whereas $\dfrac{d}{dx}$(arc sin x) is positive.

DISPLACEMENT AS A FUNCTION OF VELOCITY

The form $v\dfrac{dv}{dx}$ is useful when we are given the acceleration as a function of the velocity, and wish to relate displacement and velocity.

Example 26.6 (See p287.) A body mass 2kg is acted on by a force $(8-v)$N, where vm s^{-1} is the velocity of the body. Find the displacement of the body when the velocity is 4m s^{-1} if the body was initially at rest.

Using acceleration as $v\dfrac{dv}{dx}$,

$$8-v=2v\frac{dv}{dx}$$

$$x=\int\frac{2v}{8-v}dv$$

$$=\int\left(-2+\frac{16}{8-v}\right)dv$$

$x=-2v-16\ln(8-v)+C$. When $x=0$, $v=0$, so $C=16\ln 8$ and

$$x=16\ln\left\{\frac{8}{(8-v)}\right\}-2v$$

The displacement when $v=4$ is $16\ln 2-8$, about 3.1m.

POWER

When a body is travelling in a straight line with velocity v under a force F, the power exerted (or developed) by the force is Fv. We are probably most familiar with this term when considering cars, where we know the power P, and can deduce that the propulsive force is P/v. In practice, the power produced by the engine of a car depends on the gearing, and also on the speed of the car at that instant, but may be nearly constant over a certain range of velocities. The SI unit of power is the watt (W), which is inconveniently small, so that the kilowatt (kW) is generally used. Note carefully whether W or kW is used in examinations.

Example 26.7 A particle mass m, initially at rest, is acted on by a force F, constant in direction, whose magnitude at time t is $k\left(1+\dfrac{t}{T}\right)$, $0\leqslant t\leqslant T$. Find the power developed by F at any time t, $t\leqslant T$.

Since $F=ma$,

$$a=\frac{k}{m}\left(1+\frac{t}{T}\right)$$

$$v=\frac{k}{m}\left(t+\frac{t^2}{2T}\right),$$

the constant of integration being 0 since $v=0$ when $t=0$.

The power is Fv, i.e.

$$=\frac{k^2}{m}\left(1+\frac{t}{T}\right)\left(t+\frac{t^2}{2T}\right)$$

$$=\frac{k^2t}{m}\left(1+\frac{3t}{2T}+\frac{t^2}{2T^2}\right)$$

Example 26.8 A car mass 500kg has an engine that works at a constant rate of 40kW. If there is a constant resistance to motion of 1000N, find the time taken accelerating from 20 to 30m s^{-1}.

The propulsive force produced by the engine is $\dfrac{40\,000}{v}$N, so that $F=ma$ gives

$$\frac{40\,000}{v}-1000=500\frac{dv}{dt}$$

i.e.
$$\frac{80-2v}{v}=\frac{dv}{dt}$$

$$2t=\int_{20}^{30}\frac{v\,dv}{40-v}$$

$$=\int_{20}^{30}\left(-1+\frac{40}{40-v}\right)dv$$

$$=\int_{20}^{30}(-v-40\ln(40-v))dv$$

$$=40\ln 2-10,$$
$$\simeq 17.7\text{ s}$$

QUESTIONS

1 A car is travelling with constant acceleration 0.8m s^{-2}. If initially its velocity is 10m s^{-1}, find its velocity 5 s later and the distance travelled in that time.

2 A car travelling with constant acceleration covers a certain distance in 2 minutes. Its average speed over the distance is 60km h^{-1}, and during the second minute it travels 1200m. Find the speed at the beginning of the interval and the acceleration.

3 A train starting from rest is uniformly accelerated until its speed is 45km h^{-1}. The brakes are then applied to give a retardation twice as great as the acceleration, until its speed is 30km h^{-1}. If the distance

travelled during the retardation is 1km, find the distance travelled while accelerating.

4 A stone is projected vertically upwards with a velocity of 24.5m s^{-1} from a point 49m above the ground. Taking the acceleration due to gravity as 9.8m s^{-2}, find how long it takes to reach the ground.

5 A train starting from rest at station A accelerates uniformly until it reaches its maximum velocity of 120km h^{-1}. It moves at this constant velocity for some time, then retards uniformly until it comes to rest at station B. Given that the acceleration and retardation are numerically equal, that the distance AB is 10km and that the times of acceleration, constant velocity and retardation are numerically equal, find the time taken to go from station to station.

6 A body mass 1kg initially at rest as acted on at time t seconds by a force $(40-12t^2)$N. Find the velocity of the body after 2 s, and the displacement at that time. Find also how far the body moves in the third second of motion.

7 A lorry mass 6000kg is travelling on level road. There is a constant resistance to motion of 2000N, and the engine of the lorry exerts a force in the first minute of $50(100-t)$N, where t seconds is the time since the motion started. Find the velocity after the lorry has been travelling for 30 s, and the displacement of the lorry at that time.

8 A particle moves in a straight line so that the acceleration, at time t seconds, is $(4+6t)$m s^{-2}. If the particle travels 15m between $t=1$ and $t=2$, find the initial velocity.

9 The acceleration of a particle moving in a straight line is $(2-v)$m s^{-2}, where vm s^{-1} is the velocity of the particle. If the particle is initially at rest, find the velocity and the displacement after 1 s.

10 A particle mass 4kg initially at rest is acted on by a force $(16-v^2)$N, where vm s^{-1} is the velocity of the particle. Find the velocity of the particle after 1 s, and the limiting velocity.

11 A particle mass 4kg initially at rest is acted on by a force $(16-v^2)$N, where vm s^{-1} is the velocity of the particle. Find the displacement when the velocity is
(a)   2m s^{-1}
(b)   3.9m s^{-1}.

12 A particle at rest is acted on by a force that produces an acceleration in a constant direction of $2(1+x)$m s^{-2}, xm being the displacement from O. Find
(a)   the velocity when the particle is 8m from O,
(b)   the displacement when the velocity is 4m s^{-1}.

13 A particle unit mass is thrown vertically upwards with initial velocity U. The air exerts a resistance kv^2, where v is the velocity of the particle. Show that the greatest height reached by the particle is
$$\frac{1}{2k}\ln\left(1+\frac{ku^2}{g}\right).$$

14 A body mass 4kg initially at rest is acted on by a force F which increases uniformly from 20 N to 40 N in 10 s, the direction of F

remaining constant. Find the velocity attained by the body after 5 s and after 10 s, and the power developed by F at each of those times.

15   A car's engine exerts a constant power of 40kW, which gives the car a maximum speed on level ground of 50m s^{-1}. If the resistances to motion are proportional to the square of the speed, find the resistance when the speed is 25m s^{-1}.

16   If the mass of the car in question 15 is 800kg, show that the maximum speed the car can attain up a hill of 1 in 10 is about 10m s^{-1}.

17   A car mass 800kg has a maximum speed of 40m s^{-1} when moving along a level road, with the engine working at 60kW. Find the resistance to motion of the car.

   If the resistance varies as the square of the speed at which the car is travelling, find the power developed by the engine when the car is

   (a)   travelling along a level road at a constant speed of 30m s^{-1};

   (b)   travelling along the same road at 20m s^{-1}, and accelerating at 0.5m s^{-2}

   (c)   travelling at a constant speed of 20m s^{-1} up an incline of 1 in 100.

18   A car has a maximum speed on level road of 60m s^{-1} when the engine exerts P watts, and it can coast (i.e. run down at constant speed without power) down an incline of 1 in 20 at the same speed. If the engine exerts the same constant power P watts, how far would the car travel while accelerating on level ground from 30m s^{-1} to 45m s^{-1} the resistance to motion being proportional to the square of the speed?

USE OF VECTOR NOTATION

CONTENTS

NOTES

If $\mathbf{r}$ is the displacement vector of a point P relative to an origin O,

the **velocity** $\mathbf{v}$ of P is $\dfrac{d\mathbf{r}}{dt}$

the **acceleration** $\mathbf{a}$ of P is $\dfrac{d\mathbf{v}}{dt}$

If $\mathbf{v_A}$ is the velocity of A, $\mathbf{v_B}$ the velocity of B, the **velocity of B relative to A** is $\mathbf{v_B} - \mathbf{v_A}$.
The **momentum** of a body mass m, velocity $\mathbf{v}$ is $m\mathbf{v}$;
the **impulse** exerted by a force $\mathbf{F}$ acting over the interval $t=0$ to $t=T$ is $\int_0^T \mathbf{F}dt$;
the **kinetic energy** of the body is $\frac{1}{2}mv^2$, and the **work done** by the force $\mathbf{F}$ is $\int_0^T \mathbf{F.v}dt$,
which is equal to $\frac{1}{2}mv^2 - \frac{1}{2}mu^2$. The **power** exerted by the force is $\mathbf{F.v}$.

UNITS

The SI unit of **momentum** is newton seconds (N s),
of **kinetic energy** and **work** is joules (J),
of **power** is watts (W).

DISPLACEMENT, VELOCITY, ACCELERATION

When considering the displacement of a particle in two or three dimensions, the great advantage of vector algebra is apparent. If a particle is at a point P, coordinates (x, y), then its position vector $\mathbf{r}$ relative to the origin O can be written $x\mathbf{i}+y\mathbf{j}$ or $\begin{pmatrix} x \\ y \end{pmatrix}$ if we are working in two dimensions, $x\mathbf{i}+y\mathbf{j}+z\mathbf{k}$ or $\begin{pmatrix} x \\ y \\ z \end{pmatrix}$ in three dimensions.

The velocity of a particle is merely $\dfrac{d\mathbf{r}}{dt}$, from the definition, and the acceleration likewise is $\dfrac{d\mathbf{v}}{dt}$. If $\mathbf{r}$ is a function of t, we can obtain the velocity of the particle and its acceleration.

Example 27.1 A particle moves in a plane so that its position vector **r** at time t is $t^3\mathbf{i}+t^2\mathbf{j}$. Find the velocity and acceleration of the particle at time t, and the speed after 5 s.

Since $\mathbf{r}=t^3\mathbf{i}+t^2\mathbf{j}$

$$\mathbf{v}=\frac{d\mathbf{r}}{dt}=3t^2\mathbf{i}+2t\mathbf{j},$$

$$\mathbf{a}=\frac{d\mathbf{v}}{dt}=6t\mathbf{i}+2\mathbf{j}.$$

The speed v is the magnitude of the velocity, so after 5 s $\mathbf{v}=75\mathbf{i}+10\mathbf{j}$, and the speed is $\sqrt{(75^2+10^2)}$, about 75.7. The units are almost invariably the SI units: metres, seconds, newtons etc., so the speed is 75.7m s^{-1}.

Example 27.2 A particle mass 2kg has position vector **r** at time t seconds given in metres by $\mathbf{r}=t^4\mathbf{i}+6t\mathbf{j}$. Find the force acting on the particle after 10 s.

Since $\mathbf{v}=\dfrac{d\mathbf{r}}{dt}$ $\mathbf{v}=4t^3\mathbf{i}+6\mathbf{j}$,

Since $\mathbf{a}=\dfrac{d\mathbf{v}}{dt}$ $\mathbf{a}=12t^2\mathbf{i}$

Since $\mathbf{F}=m\mathbf{a}$ $\mathbf{F}=m(12t^2)\mathbf{i}$
                                            $=2400\mathbf{i}$

The force is 2400 N, along the x-axis.

Many examination boards at present restrict their applications of vectors to two dimensions, but really three dimensions present no extra difficulty.

Example 27.3 A particle moves so that its displacement **r** at time t is given in metres by $\frac{1}{3}t^3\mathbf{i}+5t^2\mathbf{j}+22t\mathbf{k}$. Find the acceleration of the particle at time t, and the speed of the particle after 2 seconds.

Since $\mathbf{r}=\frac{1}{3}t^3\mathbf{i}+5t^2\mathbf{j}+22t\mathbf{k}$,
          $\mathbf{v}=t^2\mathbf{i}+10t\mathbf{j}+22\mathbf{k}$
          $\mathbf{a}=2t\mathbf{i}+10\mathbf{j}$

Since $v=|\mathbf{v}|$, and the velocity after 2 s is $4\mathbf{i}+20\mathbf{j}+22\mathbf{k}$, the speed is $\sqrt{(4^2+20^2+22^2)}$, i.e. 30m s^{-1}.

RELATIVE VELOCITY

All displacements determine the position of a particle relative to some one other point, so that if $\mathbf{r}_A$ is the displacement of a particle A relative to an origin, and $\mathbf{r}_B$ is the displacement of a particle B relative to that origin, the displacement of B relative A is $\mathbf{r}_B - \mathbf{r}_A$, hence the velocity of B relative to A is

$$\frac{d}{dt}(\mathbf{r}_B - \mathbf{r}_A), \text{ i.e. } \mathbf{v}_B - \mathbf{v}_A$$

Example 27.4 Particles A and B are initially at points with position vectors $2\mathbf{i}$ and $\mathbf{j}$ respectively. A moves with constant velocity $(\mathbf{i}+\mathbf{j})$; B with constant velocity $(2\mathbf{i}-\mathbf{j})$. Find the velocity of B relative to A, and find also the least distance apart of the particles.

$$\mathbf{v}_A = \mathbf{i}+\mathbf{j}, \ \mathbf{v}_B = 2\mathbf{i}-\mathbf{j},$$

so the velocity of B relative to A is $\mathbf{v}_B - \mathbf{v}_A$, i.e. $(\mathbf{i}-2\mathbf{j})$.

At time t, $\mathbf{r}_A = 2\mathbf{i}+t(\mathbf{i}+\mathbf{j})$, $\mathbf{r}_B = \mathbf{j}+t(2\mathbf{i}-\mathbf{j})$, so that the displacement of B relative to A is $(-2\mathbf{i}+\mathbf{j})+t(\mathbf{i}-2\mathbf{j})$, i.e. $(-2+t)\mathbf{i}+(1-2t)\mathbf{j}$. If the distance apart is d,

$$
\begin{aligned}
d^2 &= (2-t)^2 + (1-2t)^2 \\
&= 5 - 8t + 5t^2 \\
&= 5(t^2 - \tfrac{8}{5}t + \tfrac{16}{25}) + \tfrac{9}{5} \\
&= 5(t-\tfrac{4}{5})^2 + \tfrac{9}{5}
\end{aligned}
$$

Therefore the least distance apart is $\sqrt{\tfrac{9}{5}}$. If we prefer, of course, we can use calculus to find the least value of d^2, but notice that d^2 is much easier to minimize than d.

We have seen above that a force can easily be described in two or three dimensions by its components. Forces can be added by adding components, so that if forces $\mathbf{F}_1$, $\mathbf{F}_2$ are such that $\mathbf{F}_1 = 3\mathbf{i}+2\mathbf{j}$, $\mathbf{F}_2 = 5\mathbf{i}-\mathbf{j}$, $\mathbf{F}_1 + \mathbf{F}_2 = 8\mathbf{i}+\mathbf{j}$. Forces $\mathbf{F}_1$, $\mathbf{F}_2$, $\mathbf{F}_3$ passing through a point, are in equilibrium if and only if their sum is zero, so that if $\mathbf{F}_1 + \mathbf{F}_2 + \mathbf{F}_3 = 0$, $(3\mathbf{i}+2\mathbf{j}) + (5\mathbf{i}-\mathbf{j}) + \mathbf{F}_3 = 0$, $\mathbf{F}_3 = -8\mathbf{i}-\mathbf{j}$.

Example 27.5 Forces $\mathbf{F}_1$, $\mathbf{F}_2$ are such that the magnitude of $\mathbf{F}_1$ is 14 N, and it acts in the direction of the vector $3\mathbf{i}+2\mathbf{j}-6\mathbf{k}$; $\mathbf{F}_2$ is of magnitude 30 N, and acts in the direction of the vector $2\mathbf{i}+10\mathbf{j}+11\mathbf{k}$. Find the magnitude of $\mathbf{F}_1 + \mathbf{F}_2$.

The magnitude of the vector $3\mathbf{i}+2\mathbf{j}-6\mathbf{k}$ is $\sqrt{(3^3+2^2+(-6)^2)}$, i.e. 7, so that

$$
\begin{aligned}
\mathbf{F}_1 &= 2(3\mathbf{i}+2\mathbf{j}-6\mathbf{k}) \\
&= 6\mathbf{i}+4\mathbf{j}-12\mathbf{k}
\end{aligned}
$$

the magnitude of $2\mathbf{i}+10\mathbf{j}+11\mathbf{k}$ is $\sqrt{(2^2+10^2+11^2)}$, i.e. 15, so that

$$
\begin{aligned}
\mathbf{F}_2 &= 2(2\mathbf{i}+10\mathbf{j}+11\mathbf{k}) \\
&= 4\mathbf{i}+20\mathbf{j}+22\mathbf{k}
\end{aligned}
$$

and $\mathbf{F}_1+\mathbf{F}_2=10\mathbf{i}+24\mathbf{j}+10\mathbf{k}$, so the magnitude of $\mathbf{F}_1+\mathbf{F}_2$ is $\sqrt{(10^2+24^2+10^2)}$, i.e. 776; about 27.9 N.

APPLICATION OF SCALAR PRODUCT

The scalar product enables us to find the angle between two vectors, and this now allows us to compare directions of vectors.

Example 27.6 The position vector of a particle at time t is $\sin t\mathbf{i}+\cos t\mathbf{j}$. Show that its acceleration is always perpendicular to its velocity.

Since $\mathbf{r}=\sin t\mathbf{i}+\cos t\mathbf{j}$,
          $\mathbf{v}=\cos t\mathbf{i}+\sin t\mathbf{j}$
and $\mathbf{a}=-\sin t\mathbf{i}-\cos t\mathbf{j}$

Two vectors are perpendicular if their scalar product is zero, and

$$\mathbf{v.a.}=(\cos t\mathbf{i}-\sin t\mathbf{j}).(-\sin t\mathbf{i}-\cos t\mathbf{j})$$
$$=0$$

so the velocity and acceleration are always perpendicular. We recognize that the particle is moving in a circle, centre the origin, radius 1.

MOMENTUM AND IMPULSE

The momentum of a body mass m, velocity $\mathbf{v}$ is $m\mathbf{v}$; the impulse $\mathbf{I}$ produced by a force $\mathbf{F}$ acting for a time T is defined as

$$\mathbf{I}=\int_0^T \mathbf{F}dt$$

Since $\mathbf{F}=m\mathbf{a}$

$$\mathbf{I}=\mathbf{F}\int_0^T m\mathbf{a}dt=m\int_0^T \frac{d\mathbf{v}}{dt}dt$$
$$=m(\mathbf{v}-\mathbf{u})$$

so that the impulse is equal to the change in momentum.

KINETIC ENERGY

The kinetic energy of a particle mass m, velocity $\mathbf{v}$ is $\frac{1}{2}mv^2$, i.e. $\frac{1}{2}m\mathbf{v.v}$. For example, the kinetic energy of a body mass 5kg, velocity $(6\mathbf{i}-8\mathbf{j})\text{m s}^{-1}$ is $\frac{1}{2}(5)(6\mathbf{i}-8\mathbf{j}).(6\mathbf{i}-8\mathbf{j})=250$ joules.

WORK AND POWER

The work done by a force $\mathbf{F}$ is defined as $\int \mathbf{F}.\mathbf{v}dt$. Since $\mathbf{F}=m\dfrac{d\mathbf{v}}{dt}$, the work done is

$$m\int \mathbf{v}.\dfrac{d\mathbf{v}}{dt}dt=m(\tfrac{1}{2}v^2-\tfrac{1}{2}u^2)$$

which is the increase in kinetic energy.

The power exerted by a force $\mathbf{F}$ is the rate of doing work, i.e.

$$\dfrac{d}{dT}\int_0^T \mathbf{F}.\mathbf{v}dt=\mathbf{F}.\mathbf{v}.$$

Example 27.7 A particle mass 5kg moves so that its position vector $\mathbf{r}$ at time t is $\mathbf{r}=\sin 2t\mathbf{i}+\cos 2t\mathbf{j}+2t\mathbf{k}$. Find (a) the momentum at time t, (b) the kinetic energy, (c) the work done on the particle in the time interval $t=0$ to $t=2$, (d) the force acting on the particle at time t, (e) the power exerted by this force.

Since $\quad \mathbf{r}=\sin 2t\mathbf{i}+\cos 2t\mathbf{j}+2t\mathbf{k}$,
$\quad\quad\quad \mathbf{v}=2\cos 2t\mathbf{i}-2\sin 2t\mathbf{j}+2\mathbf{k}$
$\quad\quad\quad \mathbf{a}=-4\sin 2t\mathbf{i}-4\cos 2t\mathbf{j}$

The momentum is $m\mathbf{v}$, so the momentum is

$\quad 5(2\cos 2t\mathbf{i}+-2\sin 2t\mathbf{j}+2\mathbf{k})$
$\quad =10(\cos 2t\mathbf{i}-\sin 2t\mathbf{j}+\mathbf{k})\text{N s}$

The kinetic energy is $\tfrac{1}{2}mv^2$, so this is

$\quad \tfrac{1}{2}(5)(2\cos 2t\mathbf{i}-2\sin 2t\mathbf{j}+2\mathbf{k}).(2\cos 2t\mathbf{i}-2\sin 2t\mathbf{j}+2\mathbf{k})$
$\quad =20 \text{ joules, which is constant}$

The work done on the particle is $\int \mathbf{F}.\mathbf{v}dt$, i.e.

$\quad \int(5)(-4\sin 2t\mathbf{i}-4\cos 2t\mathbf{j}).(2\cos 2t\mathbf{i}-2\sin 2t\mathbf{j}+2\mathbf{k})dt=0,$

as expected, since the kinetic energy is not changed.
The force acting on the particle we have already used,

$\quad \mathbf{F}=-20(\sin 2t\mathbf{i}+\cos 2t\mathbf{j})\text{N}$

and the power exerted by this force is clearly zero, since no work is being done. We can see that $\mathbf{F}.\mathbf{v}=0$, as above. The particle is describing a helix.

QUESTIONS

In all the questions in this exercise, the standard notation $\mathbf{r}$, $\mathbf{v}$, etc., is used, and the units are kg, metres and seconds.

1 A body mass 4kg has position vector $t^2\mathbf{i}+t^3\mathbf{j}$. Find its velocity and

acceleration at time t, the speed after 2 s and the force acting on the body after 10 s.

2   A body mass m has position vector $\sin t\mathbf{i}+\cos t\mathbf{j}$. Show that the body moves with constant speed under the action of a force of constant magnitude.

3   A body mass 2kg has position vector $e^{-t}\mathbf{i}+t^2\mathbf{j}+2\mathbf{k}$. Find the velocity and acceleration of the body at time t, and the force acting on the body.

4   A body mass 5kg is acted on by a force 15N in the direction of the vector $2\mathbf{i}-2\mathbf{j}+\mathbf{k}$. If the body is initially at the origin and has velocity $4\mathbf{k}$, find the displacement of the body after 2 s.

5   A body mass 3kg is acted on by a force $6t\mathbf{i}-36t^2\mathbf{j}$. If the body is initially at rest, find its velocity and displacement at time t.

6   Particles A and B have constant velocities $(\mathbf{i}+\mathbf{j})$ and $(2\mathbf{i}-\mathbf{j})$ respectively. If they are initially at points position vectors $\mathbf{i}$ and $k\mathbf{j}$ respectively, find the displacement of B relative to A at time t, and find the value of k if the particles collide.

7   Particles A and B, initially at points with position vectors $2\mathbf{i}$ and $3\mathbf{j}$, have constant velocities of $(\mathbf{i}+\mathbf{j})$ and $(2\mathbf{i}-\mathbf{j})$ respectively. Find the distance apart of A and B at time t, and when this distance is least.

8   Concurrent forces $\mathbf{F}_1$, $\mathbf{F}_2$, $\mathbf{F}_3$ are described by the vectors $2\mathbf{i}+3\mathbf{j}$, $4\mathbf{i}-5\mathbf{j}$ and $6\mathbf{i}+2\mathbf{j}$ respectively. Find the magnitude of
   (a) $\mathbf{F}_1+\mathbf{F}_2$,
   (b) $\mathbf{F}_1+\mathbf{F}_2+\mathbf{F}_3$,
   (c) $2\mathbf{F}_1+\mathbf{F}_2-3\mathbf{F}_3$.

9   Concurrent forces $\mathbf{F}_1$, $\mathbf{F}_2$ and $\mathbf{F}_3$ are such that $\mathbf{F}_1=6\mathbf{i}+5\mathbf{j}$, $\mathbf{F}_2$ is in the direction of the vector $-3\mathbf{i}-4\mathbf{j}$ and $\mathbf{F}_3$ is in the direction of the vector $\mathbf{j}$. If the system of forces is in equilibrium, find the magnitude of each of the three forces.

10   A particle mass m moves under the action of a force so that its position vector $\mathbf{r}$ is $(\frac{1}{3}t^3\mathbf{i}+\frac{1}{2}t^2\mathbf{j}+t\mathbf{k})$. Find, at time t,
   (a)   the momentum of the particle,
   (b)   the kinetic energy of the particle,
   (c)   the force acting on the particle,
   (d)   the power exerted by this force.

11   A force $(-\mathbf{i}+3\mathbf{j})$N acts on a body mass 5kg. If the body initially has a velocity of $(2\mathbf{i}-\mathbf{j})$m s^{-1}, show, by considering the impulse of the force, that 5 s later the body is moving at right angles to its initial direction.

12   A body mass 3kg, velocity $(4\mathbf{i}+5\mathbf{j})$m s^{-1}, receives an impulse of $(6\mathbf{i}-9\mathbf{j})$N s. Find the cosine of the angle through which the body is deflected.

13   A body mass 2kg is acted on by a force $\mathbf{F}$, where $\mathbf{F}=12(t\mathbf{i}+\mathbf{j})$N. If the body is initially at rest, find its velocity when $t=2$, and its momentum and kinetic energy then.

14   Particles A, B and C are such that the velocity of A is $(3\mathbf{i}+8\mathbf{j})$, of B is $(-4\mathbf{i}-\mathbf{j})$. The velocity of C relative to A is in the direction $\mathbf{i}-3\mathbf{j}$; the velocity of C relative to B is at right angles to this. Find the velocity of C.

15   Forces $i+j-k$ and $2i+3j-2k$ act at points i and $i+3j$ respectively. Show that the lines of action of these forces intersect, and find the single force to which they are equivalent.

16   When a motorist is driving with velocity $4i+3j$ the wind appears to come from the direction of $-j$; when he doubles his velocity the wind appears to come from the direction $(-i-j)$. Find the true velocity of the wind.

17   Find the cosine of the angle between the forces $F_1=i+j-3k$ and $F_2=i-3j-k$.

18   The force $F=4i+2j+3k$ moves a particle along the line from the origin to the point $5i+10j+20k$. Find the work done by this force.

19   A particle mass 4kg moves from rest at the origin under the action of two forces, each magnitude 10 N. One force is parallel to $3i-4j$; the other is parallel to $4i+3j$. Find

    (a)   the acceleration of the particle,

    (b)   the velocity of the particle after 5 s,

    (c)   the increase in kinetic energy during the fifth second of motion,

    (d)   the power exerted by the forces after 5 s.

20   A body mass 1kg, velocity $(6i-12j)$m s^{-1} embeds itself in a body mass 5kg, initially at rest, so that they move away together. Assuming that momentum is conserved, find the loss of kinetic energy due to the impact.

21   The position vector r of a point P at time t is $i+tj$. Find an expression in terms of t for θ, the angle between OP and i, and hence obtain an expression for the angular speed of OP.

22   A particle mass 2kg is initially at rest. It is then acted on by a force $3t^2i+4t^3j$ for 2 s. Find

    (a)   the final momentum,

    (b)   the final kinetic energy of the body,

    (c)   the power exerted by the force at the end of 2 s.

DIRECT IMPACT

CONTENTS

NOTES

Always draw a diagram, as in Fig 28.1, marking the directions of the velocities

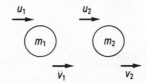

Fig 28.1

Momentum is conserved, i.e.

$$m_1\mathbf{u}_1+m_2\mathbf{u}_2=m_1\mathbf{v}_1+m_2\mathbf{v}_2$$

Newton's experimental law

velocity of separation$=e\times$velocity of approach

i.e. $v_2-v_1=e(u_1-u_2)$ *Note the signs.*

For **impact with a wall** (Fig 28.2), this becomes

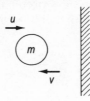

Fig 28.2

and the **impulse** given to the wall is $m(1+e)u$.

The momentum of the body parallel to the wall is not altered by the impact.

If two bodies A and B collide, then by Newton's third law, the force exerted by A on B is equal and opposite to the force exerted by B on A, while the two bodies are in contact. Since momentum is $\int_{t_1}^{t_2}\mathbf{F}dt$, the vector sum of the change in momentum of the two bodies is zero, i.e.

the momentum of the two bodies is conserved.

If the bodies have mass m_1, m_2, initial velocities $\mathbf{u}_1$, $\mathbf{u}_2$ and final velocities $\mathbf{v}_1$, $\mathbf{v}_2$, respectively (Fig 28.3), then

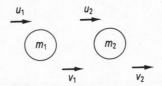

Fig 28.3

$$m_1\mathbf{u}_1+m_2\mathbf{u}_2=m_1\mathbf{v}_1+m_2\mathbf{v}_2$$

It almost invariably helps to draw a diagram and to insert the given velocities.

NEWTON'S EXPERIMENTAL LAW

Newton found by experiment that when two small spheres collide, the ratio

$$\frac{\text{velocity of separation}}{\text{velocity of approach}}$$

is independent of the velocities of the spheres before impact, providing these velocities are neither too large nor too small. This ratio is called the coefficient of restitution and denoted by e. It is an interesting property of e that for some substances e varies considerably with temperature, so that squash players warm the ball before a match, and golfers prefer to play in warm climates because the ball travels much further after impact.

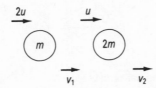

Fig 28.4

Example 28.1 Two small spheres, mass m, $2m$ are travelling in the same straight line with velocities $2\mathbf{u}$ and $\mathbf{u}$, respectively. If $e=\frac{1}{3}$, find the velocity of each sphere after the impact.
Since momentum is conserved, $2m\mathbf{u}+2m\mathbf{u}=m\mathbf{v}_1+2m\mathbf{v}_2$, $\mathbf{v}_1$ and $\mathbf{v}_2$ being the velocity of each sphere after the impact.
From Newton's experimental law,

$$\mathbf{v}_2-\mathbf{v}_1=\tfrac{1}{3}(2\mathbf{u}-\mathbf{u}) \tag{1}$$

Solving simultaneously, $\mathbf{v}_1=\frac{10}{9}\mathbf{u}$ and $\mathbf{v}_2=\frac{13}{9}\mathbf{u}$.
Always take care that the quantitites equated in equation (1) are of the same sign; the diagram should help to ensure that you have been consistent in the directions in which you suppose the bodies to be

travelling after the impact; if $\mathbf{v}_1$ is negative, it merely means the body is travelling in the opposite direction.

Example 28.2 A small body mass $3m$ moving with velocity $u\mathbf{i}$ strikes directly another small body mass $4m$, velocity $-3u\mathbf{i}$; as a result of the impact the body mass $4m$ is brought to rest. Find the coefficient of restitution e.

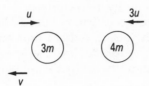

Fig 28.5

From the conservation of momentum,

$$3mu - 12mu = -3mv$$

i.e. $$v = 3u$$

From Newton's law,

$$e(4u) = v$$
$$e = \tfrac{3}{4}$$

QUESTIONS

1. A small sphere mass m_1, velocity 9m s^{-1} overtakes and collides with a similar sphere mass m_2, velocity 3m s^{-1}. After the impact the spheres continue in the same direction with velocities 5m s^{-1} and 7m s^{-1} respectively. Show that $m_1 = m_2$ and find e.

2. A particle mass $2m$, velocity $4u$ coalesces with a particle mass m, moving in the same direction with velocity u. Find the kinetic energy lost in the collision.

   If instead of coalescing, the bodies are such that $e = \tfrac{1}{2}$, find now the loss of kinetic energy.

3. Two particles A and B, each of mass m, moving in opposite directions with speeds $2u$ and u respectively, collide directly. As a result of the collision, the velocity of B is reversed but its speed is not altered. Find e and the final velocity of A.

4. A small sphere A, mass m, velocity $4u$, strikes directly another identical small sphere B moving in the same direction with velocity $2u$. After the impact the velocity of B is $3u$. Find the coefficient of restitution between A and B.

   The sphere B continues with constant velocity $3u$ until striking at right angles a smooth vertical wall, from which it rebounds and is later brought to rest by a second impact with sphere A. Find the coefficient of restitution between B and the wall.

5. Two small spheres mass m and $2m$ are moving in opposite directions

with speeds $2u$ and u. When they collide, one third of their kinetic energy is lost. Find the coefficient of restitution.

6   A small sphere mass m, velocity $3u$, strikes an identical sphere moving in the same direction with velocity u. Find the impulse of one sphere on the other, in terms of m, e and u.

7   A small sphere, mass 1kg, velocity 4im s^{-1}, strikes a similar sphere, mass 2kg, velocity im s^{-1}. If $e=0.4$, find the velocity of each sphere after the impact, the impulse exerted by each sphere on the other, and the loss of kinetic energy in the impact.

8   Two small spheres, masses m, $2m$, are fixed one to each end of a light inextensible string. The string is held taut and horizontal, with its mid-point fixed, then the spheres are released from rest. If $e=\frac{1}{2}$, show that one sphere is brought to rest by the first impact, and the other by the second impact.

PROJECTILES

CONTENTS

NOTES

If a particle is projected from a point O with velocity U at an angle θ above the horizontal,

the **time** taken to return to the horizontal level of O is $\dfrac{2U\sin\theta}{g}$

the **range** on the horizontal plane through O is $\dfrac{U^2\sin 2\theta}{g}$

the **greatest height** attained is $\dfrac{U^2\sin^2\theta}{2g}$

The **range** on the horizontal plane through O **is a maximum** if $\theta=\pi/4$.
The range on a **plane through O inclined at angle** α above the horizontal is

$$\frac{U^2}{g\cos^2\alpha}[\sin(2\theta+\alpha)-\sin\alpha]$$

the **greatest range** on this plane is

$$\frac{U^2}{g\cos^2\alpha}(1-\sin\alpha)$$

Referred to axes through O, the equation of the **trajectory** is

$$y=x\tan\theta-\frac{1}{2}\left(\frac{gx^2\sec^2\theta}{U^2}\right)$$

TIME OF FLIGHT

If a particle is projected from a point O with velocity U at an angle θ above the horizontal, the vertical component of the velocity is $U\sin\theta$; the rate of change of velocity is $-g$, the acceleration due to gravity, so the time taken to reach the highest point on the path is $U\sin\theta/g$, and the time taken to return to the horizontal level of O is $2U\sin\theta/g$.

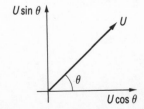

Fig 29.1

RANGE ON A HORIZONTAL PLANE

Since the horizontal component of the velocity is constant, $U\cos\theta$, the horizontal distance travelled is

$$\left\{\frac{2U\sin\theta}{g}\right\}U\cos\theta=\frac{U^2\sin 2\theta}{g}.$$

MAXIMUM RANGE ON A HORIZONTAL PLANE

The greatest value of $\sin 2\theta$ is 1, when θ is 45° or $\pi/4$ rad, so that the greatest range on a horizontal plane is U^2/g.

To find the angle of elevation required to attain a given horizontal range less than the maximum, say a range of kU^2/g.

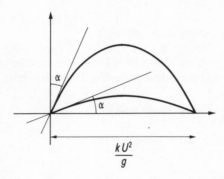

Fig 29.2

$$\frac{kU^2}{g}=\frac{U^2\sin 2\theta}{g}$$

$\therefore\quad \sin 2\theta=k$

$2\theta=\alpha$ or $\pi-\alpha$, where $\sin\alpha=k$

Thus $\quad\theta=\dfrac{\alpha}{2}$ or $\dfrac{\pi}{2}-\dfrac{\alpha}{2}.$

We see that there are two angles of elevation either of which will attain a given range.

GREATEST HEIGHT

Since the vertical component of the initial velocity is $U\sin\theta$, the greatest height attained is $U^2\sin^2\theta/2g$. This can be found by using the constant acceleration formula,

$$v^2=u^2+2as$$
$$0=(U\sin\theta)^2-2gs$$
$\therefore\qquad s=U^2\sin^2\theta/2g$

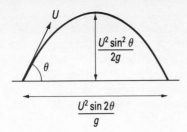

Fig 29.3

Example 29.1 A vertical wall height h stands on horizontal ground. A particle is projected in a vertical plane at right angles to the wall from a point on the ground distance d from the wall so that it just clears the wall at the highest point of its trajectory. Find the angle above the horizontal at which the particle was projected and the velocity of projection.

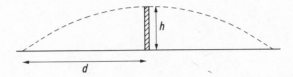

Fig 29.4

Since the particle just clears the wall at the highest point on its path, d is half the range on the horizontal plane,

$$d = \frac{V^2}{2g} \sin 2\theta$$

$$gd = V^2 \sin\theta\cos\theta \tag{1}$$

Since h is the greatest height attained,

$$2gh = V^2 \sin^2\theta \tag{2}$$

Dividing (2) by (1) gives $\tan\theta = 2h/d$, so the angle above the horizontal is arc tan $(h/2d)$.

To find V, use a right-angled triangle as shown in Fig 29.5 to find

$$\sin\theta = \frac{h}{\sqrt{(4h^2+d^2)}}, \text{ so}$$

$$V^2 = 2gh\,\frac{(4h^2+d^2)}{4h^2}$$

$$V = \sqrt{\left(\frac{g(4h^2+d^2)}{2h}\right)}$$

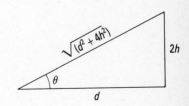

Fig 29.5

IMPACT WITH A VERTICAL WALL

If a particle is projected so that it bounces off a smooth vertical wall, the vertical velocity of the particle is unaltered, so that the time of flight is not affected by the impact. This of course is true however many impacts there are with vertical surfaces. The horizontal component of velocity is reversed and reduced by a factor e.

Example 29.2 A particle is projected with speed U from a point O distance d from a smooth vertical wall. The particle returns to O after bouncing on the wall. Find the angle α above the horizontal at which the particle was projected.

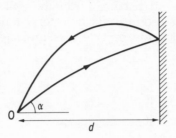

Fig 29.6

The vertical motion is not affected by the impact with the wall so the time of flight is $2U \sin \alpha/g$. The time to reach the wall is $d/(U \cos \alpha)$, and the time to return to O from the wall is $d/(eU \cos \alpha)$, so that if it returns to O,

$$\frac{2U \sin \alpha}{g} = \frac{d}{U \cos \alpha} + \frac{d}{eU \cos \alpha}$$

i.e. $$2U \sin \alpha \cos \alpha = \frac{gd}{U}\left(1 + \frac{1}{e}\right)$$

∴ $$\sin 2\alpha = \frac{gd}{U^2}\left(1 + \frac{1}{e}\right)$$

i.e. $$\alpha = \frac{1}{2} \arcsin\left[\frac{gd}{U^2}\left(1 + \frac{1}{e}\right)\right]$$

There will, of course, be two possible values for α

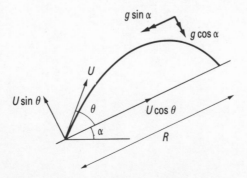

Fig 29.7

RANGE ON AN INCLINED PLANE

When considering motion relative to an inclined plane it is usually advisable to consider the velocity and acceleration along and at right angles to the plane.

Considering motion perpendicular to the plane, the initial velocity in that direction is $U \sin \theta$, the acceleration is $-g \cos \alpha$, so that the time until the velocity perpendicular to the plane is zero is $U \sin \theta / g \cos \alpha$, and the time of flight until the particle returns to the plane is $2U \sin \theta / (g \cos \alpha)$. Since $s = ut + \frac{1}{2}at^2$, as the acceleration is constant, the range R along the plane is

$$U \cos \theta \times \frac{2U \sin \theta}{g \cos \alpha} - \tfrac{1}{2}(g \sin \alpha)\frac{4U^2 \sin^2 \theta}{g^2 \cos^2 \alpha}$$

i.e.
$$R = \frac{2U^2 \sin \theta \, (\cos \theta \cos \alpha - \sin \theta \sin \alpha)}{g \cos^2 \alpha}$$

$$= \frac{2U^2 \sin \theta \cos(\theta + \alpha)}{g \cos^2 \alpha}$$

$$= \frac{U^2}{g \cos^2 \alpha}[\sin(2\theta + \alpha) - \sin \alpha]$$

As θ varies, the greatest value of $\sin 2\theta$ is 1, so that the greatest range on a plane inclined at an angle above the horizontal is $\frac{U^2}{g \cos^2 \alpha}(1 - \sin \alpha)$. Replacing α by $-\alpha$, the greatest range down a plane inclined at angle α to the horizontal is

$$\frac{U^2}{g \cos^2 \alpha}(1 + \sin \alpha)$$

PARTICLE BOUNCING ON AN INCLINED PLANE

If an elastic particle is projected so that it bounces on an inclined plane, the velocity along the line of the plane is unaltered by the impact with the plane; the velocity perpendicular to the plane is reversed and reduced by a factor e, so that the time of flight between successive bounces is reduced by a factor e.

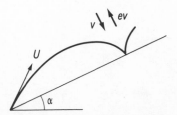

Fig 29.8

Example 29.3 A ball is projected with speed U from a point O on a smooth plane inclined at an angle α above the horizontal, so that it bounces on the plane and returns to O on the *nth* bounce. Find the tangent of the angle made by the initial velocity of the ball and the plane.

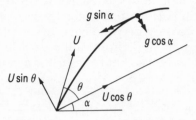

Fig 29.9

If θ is the angle between the direction of projection and the plane, the time until the first bounce is $2U \sin \theta / g \cos \alpha$, so the time until the nth bounce is

$$\frac{2U \sin \theta}{g \cos \alpha} + \frac{2eU \sin \theta}{g \cos \alpha} + \frac{2e^2 U \sin \theta}{g \cos \alpha} \cdots$$

$$= \frac{2U \sin \theta}{g \cos \alpha} (1 + e + e^2 \ldots e^{n-1})$$

$$= \frac{2U \sin \theta}{g \cos \alpha} \left(\frac{1 - e^n}{1 - e}\right)$$

In this time the ball must have returned to O, and we could use $s = ut + \frac{1}{2}at^2$ to find an equation relating u to t. It is easier though to see that if $s = 0$, $t = 2u/(-a)$, so that since along the plane $u = U \cos \theta$ and $a = -g \sin \alpha$, $t = 2U \cos \theta / g \sin \alpha$ thus

$$\frac{2U \cos \theta}{g \sin \alpha} = \frac{2U \sin \theta}{g \cos \alpha} \left(\frac{1 - e^n}{1 - e}\right)$$

$$\tan \theta = \cot \alpha \left(\frac{1 - e}{1 - e^n}\right)$$

EQUATION OF THE TRAJECTORY

Taking the point of projection as the origin of coordinates, considering the horizontal distance at time t, $x = U \cos \theta \, t$, and considering the vertical distance, $y = U \sin \theta \, t - \frac{1}{2}gt^2$. Eliminating t between these equations we have

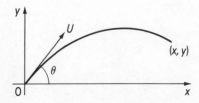

Fig 29.10

$$y = x \tan \theta - \tfrac{1}{2}\frac{gx^2}{U^2} \sec^2 \theta$$

which is the equation of the path of the particle, called the **trajectory**.

If the origin O is taken at the highest point on the path, $\theta = 0$ and the equation is $y = -\tfrac{1}{2}\frac{gx^2}{U^2}$.

TO FIND THE DIRECTION OF MOTION

The gradient of a curve is dy/dx, and since at any instant the particle is travelling along the tangent to the curve at that instant, the direction of motion is given by dy/dx.

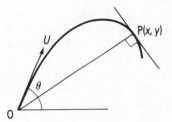

Fig 29.11

Example 29.4 A cricket ball is hit from a point O at an angle of elevation θ above the horizontal. Prove that there are two points P on the trajectory at which the direction of motion of the ball is perpendicular to OP, if and only if $\tan \theta > 2\sqrt{2}$.

The equation of the trajectory is $y = x \tan \theta - \dfrac{gx^2}{2V^2} \sec^2 \theta$

$$\therefore \qquad \frac{dy}{dx} = \tan \theta - \frac{gx}{V^2} \sec^2 \theta$$

The gradient of OP is $\dfrac{y}{x}$, i.e. $\tan \theta - \dfrac{gx}{2V^2} \sec^2 \theta$.

If the direction of motion is perpendicular to OP,

$$\left(\tan \theta - \frac{gx}{V^2} \sec^2 \theta\right)\left(\tan \theta - \frac{gx}{2V} \sec^2 \theta\right) = -1$$

i.e. $\qquad \tan^2\theta - \dfrac{3gx}{2V^2}\sec^2 \theta \tan \theta + \dfrac{g^2x^2}{2V^4} \sec^4 \theta = -1$

i.e.
$$\frac{g^2x^2}{2V^4}\sec^2\theta - \frac{3gx}{2V^2}\tan\theta + 1 = 0,$$

using $\tan^2\theta + 1 = \sec^2\theta$ and dividing by $\sec^2\theta$. This is a quadratic in x, with two real distinct roots if

$$\frac{9g^2}{4V^2}\tan^2\theta > 4\frac{g^2\sec^2\theta}{2V^4}$$

i.e. $9\tan^2\theta > 8(\tan^2\theta + 1)$

i.e. $\tan\theta > 2\sqrt{2}.$

QUESTIONS

1   A golfer drives a ball over a flat horizontal course, giving it a velocity of 35m s^{-1} when it leaves the club. What is the greatest distance he can drive the ball? At what angle to the horizontal should he strike the ball if it is to have a range of 100m, the velocity with which it leaves the club still being 35m s^{-1}? Find the greatest height attained by the ball on each of the last two paths.

2   A particle is projected with a velocity whose horizontal and vertical components are u and v respectively. Find the range on a horizontal plane through the point of projection.

3   A particle is projected from a point on a plane inclined at 30° above the horizontal. Prove that the greatest range down the plane is three times the greatest range up the plane.

4   A particle is projected with velocity V at an angle $(\alpha+\theta)$ above the horizontal from a point in a plane inclined at angle θ above the horizontal. Show that the particle strikes the plane at right angles if $2\tan\alpha = \cot\theta$.

5   A ball is thrown with speed 14m s^{-1} from a window 25m above a level horizontal playground and lands 10m from the point vertically below the point of projection. Find the tangent of the angle, the direction of motion the ball made initially with the horizontal, and the direction of motion, 1 s later, of the ball along each of the two possible paths. Find also the ratio of the times of flight of the ball along the two possible trajectories.

6   A ball is thrown from a point P on smooth horizontal ground towards a smooth vertical wall, distance d from P. The ball returns to P, having bounced on the wall and having bounced once on the ground. The coefficient of restitution between the ball and the wall, and between the ball and the ground is e. If the speed of projection is $\sqrt{(2gd/e)}$, find the angle above the horizontal at which the ball was projected. Sketch the two possible paths of the ball.

MOTION IN A CIRCLE

CONTENTS

A particle will not describe a circle unless it has an acceleration towards the centre of the circle of v^2/r, or $r\omega^2$. Therefore there must be a central force of $\boldsymbol{mv^2/r}$.

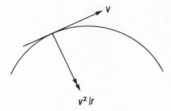

Fig 30.1

CONICAL PENDULUM

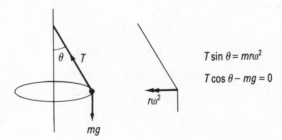

$$T \sin \theta = mr\omega^2$$
$$T \cos \theta - mg = 0$$

Fig 30.2

MOTION IN A VERTICAL CIRCLE

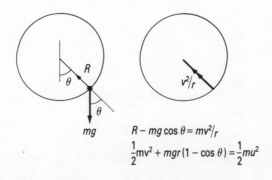

$$R - mg \cos \theta = mv^2/r$$
$$\frac{1}{2}mv^2 + mgr(1 - \cos \theta) = \frac{1}{2}mu^2$$

Fig 30.3

ACCELERATION TOWARDS THE CENTRE

Newton's first law observes that a particle stays at rest or travels in a straight line with constant velocity unless a force acts on that particle, so a particle cannot travel in any curved path unless there is an acceleration at right angles to the direction of motion. If the particle is travelling around a circle with constant speed there will not be a tangential force but there must always be a force towards the centre to produce the central acceleration. There are many ways of proving that this central acceleration is v^2/r (or $r\omega^2$), where ω is the angular speed. The one below is often found easier than the others, and uses parameters to describe the circular path.

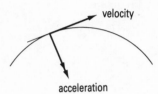

Fig 30.4

USE OF PARAMETERS

The position of any particle travelling round a circle centre the origin radius r, is given by

$$x=r \cos \theta, \; y=r \sin \theta$$

Differentiating with respect to time,

$$\dot{x}=-r \sin \theta \dot{\theta}, \text{ and } \dot{y}=r \cos \theta \dot{\theta}$$

$$\ddot{x}=-r \cos \theta \dot{\theta}^2-r \sin \theta \ddot{\theta}$$

and
$$\ddot{y}=-r \sin \theta \dot{\theta}^2+r \cos \theta \ddot{\theta}$$

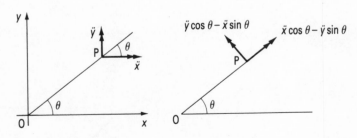

Fig 30.4

The components of acceleration outwards in the direction of OP (Fig 30.5) are

$$\ddot{x} \cos \theta+\ddot{y} \sin \theta$$

i.e. $(-r \cos \theta \dot{\theta}^2-r \sin \theta \ddot{\theta}) \cos \theta+(-r \sin \theta \dot{\theta}^2+r \cos \theta \ddot{\theta}) \sin \theta$

i.e. $-r\dot{\theta}^2$

so the acceleration is $r\dot{\theta}^2$ or $r\omega^2$ towards the centre.

The components of acceleration along the tangent in the direction of θ increasing are

$$\ddot{y}\cos\theta - \ddot{x}\sin\theta$$

i.e. $(-r\sin\theta\dot{\theta}^2 + r\cos\theta\ddot{\theta})\cos\theta - (-r\cos\theta\dot{\theta}^2 - r\sin\theta\ddot{\theta})\sin\theta$

i.e. $r\ddot{\theta}$

so that the acceleration along the tangent is $r\ddot{\theta}$ or $r\dfrac{d\omega}{dt}$.

If the particle is describing a circle with constant angular speed ω, the acceleration along the tangent is zero, as we expect.

MOTION IN A HORIZONTAL CIRCLE

If a particle mass m is to move in a horizontal circle, there must be a force towards the centre mv^2/r. This force may be provided by a string, a rod, or by friction or other forces.

Example 30.1 A toffee mass 12g is placed on a rough gramophone turntable. When the turntable is rotating at 45rpm the toffee describes a circle radius 15cm. Find the frictional force exerted by the turntable on the toffee.

Since the toffee describes a circle, there must be a force towards the centre $mr\omega^2$. Express all the quantities in SI units. 15cm=0.15m, 45rpm=1.5 rad s^{-1}, 12g=0.012kg, so that the frictional force $mr\omega^2$ is

$$(0.012)(0.15)(1.5\pi)^2 \text{ newtons, about } 0.04\text{ N}$$

Example 30.2 A particle is suspended by a string, length 30cm, one end of which is attached to a fixed point O. The particle describes a circle in a horizontal plane, with constant angular speed ω. Find the distance of the plane of the circle below O if (a) ω=10 rad s^{-1}, (b) ω=150 rad s^{-1}.

Fig 30.6

Since the particle, mass m, describes a circle in a horizontal plane, there is no vertical acceleration,

$$T \cos \theta - mg = 0 \qquad (1)$$

Since it describes a horizontal circle, there is an acceleration towards the centre of $(0.3 \sin \theta)\omega^2$

$$T \sin \theta = m(0.3 \sin \theta)\omega^2 \qquad (2)$$

Dividing (1) by (2),

$$\cos \theta = \frac{g}{0.3\omega^2}$$

When $\omega = 10$, $\cos \theta = g/30$, and the distance below O is $0.3 \times g/30$m, about 10cm, whereas when $\omega = 50$, $\cos \theta = g/750$, and the distance below O is $0.3 \times g/750$m, about 0.4cm.

Thus as the angular speed increases, the plane of the circle rises closer to the level of the fixed point O.

MOTION IN A VERTICAL CIRCLE

A particle is attached to one end of a string length r, the other end of which is attached at a fixed point O, and the particle is made to describe a circle in a horizontal plane by being given a horizontal velocity u when hanging with the string vertical. If subsequently the string makes an angle θ with the vertical and the velocity of the particle then is v, the force towards the centre now is $T - mg \cos \theta$, so $T - mg \cos \theta = mv^2/r$. But by the conservation of energy,

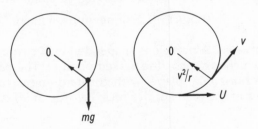

Fig 30.7

$$\tfrac{1}{2}mu^2 = \tfrac{1}{2}mv^2 + mgr(1 - \cos \theta)$$

i.e. $$v^2 = u^2 - 2gr(1 - \cos \theta)$$

so that $$T - mg \cos \theta = \frac{mu^2}{r} - 2mg(1 - \cos \theta)$$

i.e. $$T = \frac{mu^2}{r} - mg(2 - 3 \cos \theta)$$

As θ increases, T decreases, and the least value of T occurs when $\theta = \pi$, and is $mu^2/r - 5mg$. If the particle is to describe a complete circle at the end of a string, T must never be zero, so that $u^2 \geqslant 5gr$.

It may be that the initial velocity u is small, and then the particle will only rise above the horizontal level of O if $\frac{1}{2}mu^2 > mgr$, i.e. $u^2 > 2gr$. Otherwise it will oscillate about the vertical position through O. Thus we have

(1) $u^2 \leqslant 2gr$, particle oscillates in a vertical plane,

(2) $2gr < u^2 < 5gr$, particle rises above the level of O, the string becomes slack, then the particle travels in a parabola until the string is taut again, or

(3) $5gr \leqslant u^2$, the particle describes a vertical circle, centre O.

If the particle is attached to a rod, so that T can be both positive and negative, then u need only be sufficiently large for there to be enough kinetic energy for the particle to rise a vertical distance of $2r$, i.e. $\frac{1}{2}mu^2 \geqslant 2mgr$, $u^2 \geqslant 4gr$ for the particle to describe a circle.

Example 30.3 A smooth circular tube is fixed in a vertical plane. A small bead, mass m, is released from the highest point in the tube and slides around the tube. Find where the contact force between the bead and the tube vanishes.

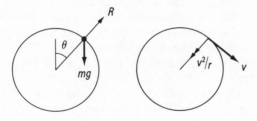

Fig 30.8

Since the bead is 'released' at the top point, we can assume the initial velocity is zero. If V is the velocity when the radius through the particle makes an angle of θ with the upward vertical (Fig 30.8), the conservation of energy gives

$$mga = \tfrac{1}{2}mV^2 + mga \cos \theta,$$

a being the radius of the circle

i.e. $V^2 = 2ga(1 - \cos \theta)$ (1)

Considering the forces along the radius

$$mg \cos \theta - R = mV^2/a$$

so that when $R = 0$, $V^2 = ga \cos \theta$.
Substituting in (1), $ga \cos \theta = 2ga(1 - \cos \theta)$

$$\theta = \text{arc} \cos (\tfrac{2}{3})$$

1   A particle mass M is attached to one end of a light inextensible string length l, the other end of which is fixed. The particle describes a horizontal circle with constant speed at m revolutions per second. Find the tension in the string and the angle the string makes with the vertical.

2   A particle mass m is suspended from a point O by a light inextensible string length $5a$, and is made to describe a horizontal circle radius $3a$ on a smooth horizontal table with a constant speed of $\sqrt{(2ga)}$. Find the reaction between the table and the particle.

   If the speed of the particle is increased slowly until the particle is about to leave the surface of the table, what is then the speed of the particle?

3   A string length 0.4m passes through a fixed smooth ring, and joins two bodies mass m and $2m$. The heavier mass hangs vertically below the ring, and the lighter describes a horizontal circle 0.1m below the ring. Find the distance below the ring of the heavier of the two masses.

4   A particle P, mass m, is suspended from a fixed point O by a light inextensible string length l. The particle hangs freely in equilibrium, and is then given a horizontal speed of $\sqrt{(3lg)}$. Find the height of P above O when the string becomes slack.

5   A particle is suspended from a fixed point O by a string length 1 metre. It is hit by an identical particle with horizontal velocity 10.5m s^{-1}, and receives just enough velocity for it to describe a vertical circle centre O. Find the coefficient of restitution between the two particles.

ELASTICITY

CONTENTS

HOOKE'S LAW

Tension is proportional to extension,

i.e. $T = kx$,

where k is the **spring constant** (units N m^{-1})

or $T = \dfrac{\lambda}{a}x$,

where λ is the **modulus of elasticity** (units N)

The **work done** in stretching an elastic string so that the extension increases from x_1 to x_2 is $\displaystyle\int_{x_1}^{x_2} kx\,\mathrm{d}x$

$$= \tfrac{1}{2}k(x_2^2 - x_1^2)$$

or $\dfrac{1}{2}\dfrac{\lambda}{a}(x_2^2 - x_1^2)$.

This is sometimes called the **elastic energy** in the string.

When a particle is oscillating while fixed to an elastic string, the equation of motion is of the form

$$\frac{\mathrm{d}^2x}{\mathrm{d}t^2} = -n^2x$$

for some n. This is **simple harmonic motion**, period $2\pi/n$.

HOOKE'S LAW

It can be shown by experiment that for any given spring the tension T is proportional to the extension x, i.e.

$$T = kx$$

where the constant k is called the spring constant. The law holds for a range of values of x, and for an elastic spring is also true when the spring is compressed, i.e. when x is negative. Spring constant is measured in newtons/metre, dimensions MT^{-2}.

For differing lengths of the same type of spring (or elastic string) then it can be shown by experiment that if we double the unstretched length a, a given force T will produce twice as much extension, so that the law can be written

$$T = \frac{\lambda}{a} x$$

where a is the unstretched length and λ is constant for that length of elastic. λ is called the **modulus of elasticity**. Notice by checking the dimensions that λ has the units of force (dimensions MLT^{-2}), and is measured in newtons.

ENERGY IN A STRETCHED SPRING

If a spring, spring constant k, is stretched so that its extension increases from x_1 to x_2, the work done is $\int_{x_1}^{x_2} T dx$. But $T = kx$, so the work done is $\int_{x_1}^{x_2} kx dx$, i.e.

$$\text{work done} = \tfrac{1}{2}kx_2^2 - \tfrac{1}{2}kx_1^2 = \tfrac{1}{2}k(x_2 + x_1)(x_2 - x_1)$$

But the initial and final tensions T_1 and T_2 are kx_1 and kx_2, so that the work done is $\tfrac{1}{2}(T_1 + T_2)(x_2 - x_1)$, the product of the mean tension and the extra extension. This work is stored in the spring as elastic energy.

SPRINGS AND STRINGS

Take care to distinguish between elastic springs, in which $T = kx$ whether x is positive or negative, and elastic strings, which become slack (and so $T = 0$) when they are not extended.

HARDER EXAMPLES

Example 31.1 An elastic string natural length $2a$, spring constant k has its ends fixed at points A, B, a distance $4a$ apart, on a smooth horizontal table. A particle mass m is attached to the string at C, the midpoint of the string. The particle is then displaced from X to Y a distance a towards A, and released from rest. Show that in the subsequent motion the time taken for the particle first to reach the midpoint of AB is $(\pi/2)\sqrt{(m/2k)}$, and its greatest speed is $a\sqrt{(2k/m)}$.

Draw a diagram, as in Fig 31.1, mark clearly the extension in each part of the spring and the appropriate force when the particle is at a distance x from the equilibrium position, the midpoint of AB.

The extension in the right-hand half of the string is $(a - x)$, so the tension is $k(a - x)$ in the direction of x increasing; the tension in the

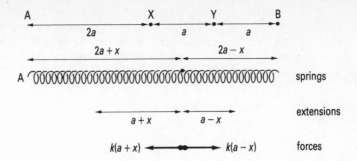

Fig 31.1

left-hand half is $(a+x)$, so the tension is $k(a+x)$ in the direction of x decreasing. Using $F=ma$,

$$k(a-x)\times k(a+x)=m\ddot{x}$$
$$\ddot{x}+\frac{2k}{m}x=0$$

This is an example of **simple harmonic motion**, and we can write down (see p262) the solution of this differential equation that has the initial condition $x=a$, $\dot{x}=0$ (since the particle is released from rest),

$$x=a\cos\left(t\sqrt{\frac{2k}{m}}\right)$$

The time that elapses before the particle first reaches the midpoint of AB is one-quarter of that for a complete oscillation, i.e. $\frac{1}{4}(2\pi)/\sqrt{(2k/m)}$

i.e.
$$\frac{\pi}{2}\sqrt{\frac{m}{2k}}.$$

To find the greatest velocity,
$v=dx/dt=a\sqrt{(2k/m)}\sin t\sqrt{(2k/m)}$
and the greatest value of this $a\sqrt{(2k/m)}$.

If we had wanted to find the time before the particle reached any other point, say that point half-way between X and Y, then using $x=a\cos t\sqrt{(2k/m)}$ we have

$$\tfrac{1}{2}a=a\cos t\sqrt{\frac{2k}{m}}$$

$$t\sqrt{\frac{2k}{m}}=\text{arc cos }(\tfrac{1}{2})$$

$$t=\frac{\pi}{3}\sqrt{\frac{m}{2k}}$$

N.B. Note the dimensions of $\sqrt{(2k/m)}$ are T^{-1}, since k has dimensions MT^{-2}, so the dimensions of $a\sqrt{(2k/m)}$ are LT^{-1}, velocity, and of $(\pi/3)\sqrt{(m/2k)}$ are T.

Example 31.2 A particle mass m is attached to one end of a light elastic string, the other end of which is attached at a fixed point O. The natural length of the string is a and the modulus of elasticity is $4mg$. If the particle is released from rest at O, show that it first comes to rest a distance $2a$ below O.

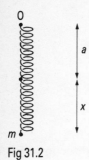

Fig 31.2

Since the velocity of the particle is 0 when it is released, and 0 when it first comes to rest, we see that the kinetic energy of the particle is zero at both instants, so that the potential energy lost is equal to the work done in stretching the string, i.e. if the particle first comes to rest when the extension is x,

$$mg(a+x)=\tfrac{1}{2}\frac{4mg}{a}x^2$$
$$2x^2-ax-a^2=0$$
$$x=a \text{ or } -\tfrac{1}{2}a$$

The particle first comes to rest when the extension is a, i.e. at a distance $2a$ below O.

QUESTIONS

1 Find the spring constant k if
   (a)   a force of 40 N produces an extension of 0.5m,
   (b)   a force of 4×10^3 N produces an extension of 2.5m.
2 Find the modulus of elasticity λ if
   (a)   a force of 40 N produces an extension of 0.5m in an elastic spring, unstretched length 2m;
   (b)   a force of 40 N produces an extension of 0.5m in an elastic spring, unstretched length 5m.
3 Find the force needed to stretch a spring, elastic constant 25 Nm^{-1}, by 4cm.
4 Find the force needed to stretch a spring unstretched length 2m, modulus of elasticity 25 N, by 4cm.
5 A light elastic string, natural length c, is attached at one end to a fixed point O. A particle mass m is attached to the other end of the string and allowed to hang freely. Find the extension in the string if the modulus of elasticity is
   (a) mg,   (b) $2mg$,   (c) $\tfrac{1}{2}mg$.
6 An elastic spring, natural length, 0.2m, spring constant 160 Nm^{-1}, is placed on a smooth horizontal table. One end of the spring is fixed at a point on the table. Find the work done when the string is stretched so that the total length increases
   (a)   from 0.2m to 0.25m,
   (b)   from 0.25m to 0.3m,
   (c)   from 0.25m to 0.5m.
7 Using the data of Example 31.2, write down the equation of motion in the form $m\ddot{x}=f(x)$, and show that the particle is executing S.H.M. about a point $\tfrac{3}{4}a$ below 0. Find the period of this motion.
8 One end of a light elastic string, natural length a, modulus of elasticity λ, is attached at a fixed point A, and a small pan, mass M, is attached to the other end so that it hangs freely in equilibrium. A small piece of putty, mass m, is then fixed underneath the pan.
   Show that the pan can now rest in equilibrium a distance $a(1+(m+M)g/\lambda)$, below A.

If the pan is slightly displaced a distance d vertically downwards from this position, find the greatest speed of the pan in the ensuing motion, and the greatest force between the putty and the pan.

9 A light elastic string, spring constant k, natural length $3a$, has its ends fixed at points A and D, a distance $3a$ apart and on the same horizontal level. Two particles, each mass m, are fixed one at each of B and C, the points of trisection of AD, and hang freely in equilibrium. Show that, if θ is the angle made by AB and CD with the horizontal, $2ka(\tan \theta - \sin \theta) = 3mg$.

FRICTION

CONTENTS

The **laws of friction** required now are:
(a)   Friction always opposes **relative** motion.
(b)   The frictional force is **just sufficient** to prevent relative motion, up to a certain maximum. This maximum frictional force is called **limiting friction**.
(c)   The limiting frictional force is proportional to the **normal reaction** between the two bodies in contact.

LAWS OF FRICTION

Experiments and observations lead us to formulate certain laws to describe the nature of the forces due to friction. The ones we need to know now are:
(a)   Friction always opposes **relative** motion.
(b)   The frictional force is just sufficient to prevent relative motion, up to a certain maximum value, called **limiting friction**.
(c)   The limiting frictional force is proportional to the **normal reaction** between the bodies in contact.

N.B. *Law* (a) Friction opposes relative motion, and so can produce motion. A parcel on the back shelf of a car will not slip backwards when the car accelerates slowly if the frictional force can keep it in place. If, however, the frictional force is not sufficient to make the parcel accelerate as rapidly as the car, the parcel will slip backwards relative to the car.

Law (c) The frictional force is proportional to the normal reaction, *not* to the weight of the body.

Example 32.1 A body mass 4kg is at rest on a rough horizontal surface, which can exert a limiting frictional force of 2N. A horizontal force F N is applied, given at time t seconds by

$$F = \tfrac{1}{10}t,$$

Find when the body starts to slip, and how far it has moved 10 seconds later.

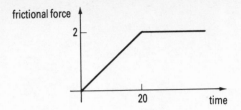

Fig 32.1

Since the maximum frictional force is 2N, for the first 20 s, i.e. until $F=2$, the frictional force is just sufficient to balance F, as shown in Fig 32.1. When $t \geqslant 20$, however, using Newton's Law,

$$4\frac{dv}{dt} = \tfrac{1}{10}t - 2$$

Integrating $v = \tfrac{1}{80}t^2 - \tfrac{1}{2}t + C$

When $t = 20$, $v = 0$, so $C = 5$,

$$v = \tfrac{1}{80}t^2 - \tfrac{1}{2}t + 5$$

Since $s = \int v\,dt$, $s = \tfrac{1}{240}t^3 - \tfrac{1}{4}t^2 + 5t + C$. When $t = 20$, $s = 0$, so $C = -\tfrac{100}{3}$ and

$$s = \tfrac{1}{240}t^3 - \tfrac{1}{4}t^2 + 5t - \tfrac{100}{3}$$

When $t = 30$, the distance travelled is 4.17m. We can check this is reasonable, for the force 'to spare' to accelerate increases from zero to 1N over 10s, so the acceleration increases from 0 to 0.25m s^{-2} over 10s. At a constant acceleration of 0.1m s^{-2} for 10s, the distance travelled would be 5m.

Example 32.2 A body weight 20N is at rest on a rough horizontal surface. A force F acts on the body. If the body is about to move under the action of a force F, magnitude 10N, find the coefficient of friction in each of the following cases: (a) F acts horizontally, (b) F acts at an angle of 30° above the horizontal, (c) F acts at an angle 30° below the horizontal.

(a)   Considering the vertical forces on the body, the normal reaction X_1 is equal to the weight of the body, 20N, so that if μ_1 is the coefficient of friction,

$$10 = 20\mu_1,$$
$$\mu_1 = 0.5$$

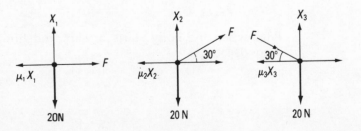

Fig 32.2

(b) Considering now the vertical forces,

$$X_2 + F \sin 30° = 20,$$
$$X_2 = 20 - 10 \times 0.5$$
$$= 15$$

so $\qquad 10 \cos 30° = 15\mu_2,$

$$\mu_2 = 0.577$$

(c) Considering in this third case the vertical forces,

$$X_3 = 20 + 10 \sin 30°$$
$$= 25$$

so $\qquad 10 \cos 30° = 25\mu_3$

$$\mu_3 = 0.346$$

Notice that in parts (b) and (c), the greater the normal reaction, the smaller the coefficient of friction necessary to produce a given horizontal force.

Example 32.3 A body mass m is at rest on a rough plane inclined at an angle α above the horizontal. Find the force F that will just move the body up the plane if (a) F acts along a line of greatest slope of the plane, (b) F acts horizontally, (c) we require the least force F, in any direction.

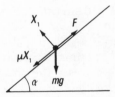 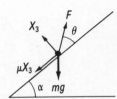

Fig 32.3

(a) The normal reaction X_1 is $mg \cos \alpha$, so the frictional force is $\mu mg \cos \alpha$. That will oppose relative motion, so will act *down* the plane. Considering then the forces along a line of greatest slope of the plane,

$$F = mg \sin \alpha + \mu mg \cos \alpha$$
i.e. $\qquad F = mg(\sin \alpha + \mu \cos \alpha)$

(b) Considering the forces normal to the plane,

$$X_2 = F \sin \alpha + mg \cos \alpha$$

so the frictional force is $\mu(F \sin \alpha + mg \cos \alpha)$, and considering the forces along a line of greatest slope of the plane,

$$F \cos \alpha = \mu(F \sin \alpha + mg \cos \alpha) + mg \sin \alpha$$

$$F(\cos \alpha - \mu \sin \alpha = \mu mg \cos \alpha + mg \sin \alpha$$

$$F = \frac{mg(\mu \cos \alpha + \sin \alpha)}{\cos \alpha - \mu \sin \alpha}$$

(c)   Suppose the force F acts at an angle of θ above the plane. Then considering the components of forces normal to the plane,

$$X_3 + F \sin \theta = mg \cos \alpha$$

so the frictional force is $\mu(mg \cos \alpha - F \sin \theta)$. Considering the forces along a line of greatest slope of the plane,

$$F \cos \theta = \mu(mg \cos \alpha - F \sin \theta) + mg \sin \alpha$$

$$F = \frac{mg(\mu \cos \alpha + \sin \alpha)}{\cos \theta + \mu \sin \theta}$$

The only variable in this expression is θ in the denominator, so that the least value of F occurs at the greatest value of $\cos \theta + \mu \sin \theta$, i.e. $\sqrt{(1 + \mu^2)}$,* the least value of F is

*See page 178

$$\frac{mg}{\sqrt{(1 + \mu^2)}} (\mu \cos \alpha + \sin \alpha)$$

QUESTIONS

1   A body mass 2kg is at rest on a rough horizontal surface, which can exert a limiting frictional force of 5N. A horizontal force F N is now applied, given at time t by $F = 10 \sin (\pi t/6)$. Find when the body starts to slip, and its velocity 2s later.

2   A rough horizontal platform is oscillating along a horizontal straight line so that its displacement x at time t is given by $x = a \cos t$. A particle is placed on the platform when it is at the position $x = a$. Show that the particle slips immediately if $g\mu < a$.

3   A particle is projected with velocity u from a point P up a line of greatest slope of a rough plane inclined at an angle α above the horizontal. If μ is the coefficient of friction between the particle and the plane, find in terms of a, g and μ how long elapses before the particle comes momentarily to rest, how far up the plane the particle travels, and how much longer elapses before the particle returns to P if $\mu < \tan \alpha$.

(More questions requiring knowledge of the laws of friction are in Questions, page 347.)

COMPOSITION AND RESOLUTION OF FORCES, MOMENTS OF A FORCE

CONTENTS

Forces are added by the **parallelogram law**, so that the resultant **R**

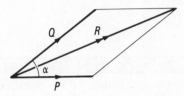

Fig 33.1

of forces **P** and **Q** inclined at an angle α has magnitude

$$\sqrt{(P^2 + Q^2 + 2PQ \cos \alpha)}$$

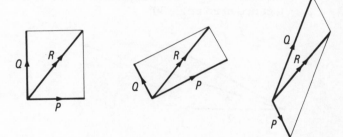

Fig 33.2

Any two forces **P** and **Q** whose vector sum is **R** can be called **components** of **R**. When these two forces are at right angles, e.g. **X** and **Y** (Fig 33.3), they are called the **resolved parts** along Ox and Oy respectively.

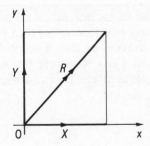

Fig 33.3

The moment of a force **F** about an axis through a point O perpendicular to the plane of **F** and O is the product of the magnitude of **F** and the perpendicular distance on to the line of action of **F**, here pF. In SI units, it is measured in newton-metres (dimensions $ML^2 T^{-2}$).

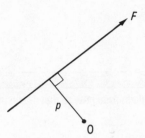

Fig 33.4

ADDITION OF FORCES

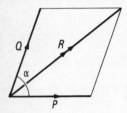

Fig 33.5

Forces being vectors, are added by the parallelogram law, so that the resultant of two forces **P** and **Q** inclined at an angle α is $\sqrt{(P^2+Q^2+2PQ \cos \alpha)}$, using the cosine formula to find the length of magnitude of their resultant.

Example 33.1 Find the magnitude of the resultant forces 2N and 3N, inclined at an angle 30°.

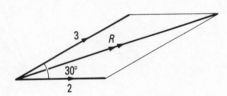

Fig 33.6

Using Fig 33.6 the magnitude of the resultant is

$$\sqrt{(2^2+3^2+2\times2\times3 \cos 30°)}, \text{ about } 4.8N.$$

COMPOSITION OF FORCES

Often, instead of adding two forces, we find it convenient to express a single force as two forces, so that a force 4N acting at 60° to the x-axis has a component $4 \cos 60°$ along the x-axis, $4 \sin 60°$ at right angles to it. This is particularly useful when we are familiar with vector notation.

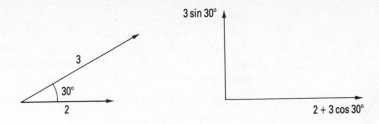

Fig 33.7

Example 33.2 Find the magnitude of the resultant of two forces 2N, 3N inclined at 30°.

Take **i** as a unit vector along the line of action of the force 2N, which can then be written 2**i** N. The other force **F** can be written (3 cos 30**i**+3 sin 30**j**) N, about (2.6**i**+1.5**j**) N. The resultant of the two forces is (4.6**i**+1.5**j**) N, and its magnitude is $\sqrt{[(4.6)^2+(1.5)^2]}$, about 4.8N as before.

MOMENT OF A FORCE

The moment of a force **F** acting about an axis through a point O perpendicular to the plane of O and **F** is defined as the product of the magnitude of **F** and the perpendicular distance p of the line of action of **F** from O, i.e. Fp, using the data in Fig 33.8. The moment measures the turning effect of the force **F** about the axis through O.

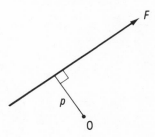

Fig 33.8

RIGID BODIES IN EQUILIBRIUM

A rigid body will only be in equilibrium if the vector sum of the forces acting on the body is zero, so that there is no acceleration of the body, and if the sum of the moments of the forces acting on the body is zero, so that there is no turning effect of those forces and the body does not rotate. Many of the examples testing understanding of this idea use a rod with one end on rough ground and the other end resting against a wall.

Example 33.3 A uniform straight rod, weight 20N, rests with one end on rough horizontal ground, the other end being against a smooth ver-

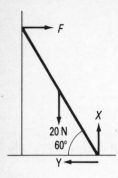

Fig 33.9

tical wall. If the rod is inclined at 60° to the horizontal, find the force exerted by the wall on the rod.

If the length of the rod is $2l$, the moment of the horizontal force **F** about the foot of the ladder is $2l \sin 60°\ F$ and the moment of the weight of the rod is $l \cos 60° \times 20$ in the opposite sense, both being in newton-metres. Since the rod is in equilibrium,

$$2l \sin 60°\ F = l \cos 60° \times 20$$

$$F = 10 \cos 60°$$

$$\approx 5.77$$

The force between the wall and the rod is 5.77N.

Considering the horizontal forces on the rod, we see that the frictional force exerted by the ground must be 5.77N, so that the rod does not move horizontally, and the normal reaction between the rod and the ground is 20N. Since the rod does not slip,

$$5.77 \leqslant 20\mu$$

The coefficient of friction must be at least 0.289.

Example 33.4 A uniform rod AB, length $2a$, weight 3N, has a light ring at one end A, which is free to slide on a rough horizontal wire. One end of a light string length a is fixed to the point B of the rod and the other end is attached at a fixed point C along the wire. Find the tension in the string, if the rod is inclined at an angle of 20° to the horizontal.

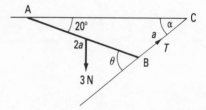

Fig 33.10

Since we do not wish to find the forces on the rod through A, take moments about A,

$$3(a \cos 20°) = T \times 2a \sin \theta$$

To find θ, we need to know angle ACB$=\alpha$. Using the sine formula,

$$\frac{a}{\sin 20°} = \frac{2a}{\sin \theta}$$

$$\sin \alpha = 2 \sin 20°$$

$$\alpha \approx 43.2°$$

Thus $\theta = 63.2°$, and

$$3 \cos 20° = T(2 \sin 63.2°)$$

$$T = 1.58$$

The tension is about 1.58N.

Using each of the two methods above, find the magnitude of the resultant of the following forces.

1 Forces 4N, 5N inclined at 70°.
2 Forces 4N, 5N inclined at 110°.
3 Forces 4N, 5N inclined at 20°.
4 The foot of a uniform ladder, weight 400N, rests on rough horizontal ground and the top of the ladder rests against a smooth vertical wall. A man, weight 800N, can stand on the top of the ladder when the ladder makes an angle of 70° with the horizontal. Find the force exerted by the wall on the ladder, and the least value of the coefficient of friction between the ladder and the ground.
5 A uniform ladder, length l, weight 80N, rests inclined at 70° to the horizontal with one end on rough horizontal ground, the other end against a smooth vertical wall. The coefficient of friction between the ladder and the ground is 0.3. A man, weight 800N, starts to climb the ladder slowly.
   (a)   Write down the normal reaction exerted by the ground on the ladder, and deduce the greatest frictional force exerted by the ground on the ladder.
   (b)   Find the horizontal force exerted by the wall on the ladder when the man has ascended a length kl of the ladder.
   (c)   Show that the ladder starts to slip just before the man reaches the top of the ladder.
6 A uniform rod, weight 100N, length 1.6m, rests inclined at 70° to the horizontal with one end on rough horizontal ground, supported also by a smooth peg 0.6m above the ground. Find the force exerted by this peg on the rod.
7 A uniform rod AB, length 1m, weight 40N, is smoothly hinged to a vertical wall at A. It is held at rest in a horizontal position by a light inextensible string attached at B and at a point C on the wall a distance 0.3m above A. Find the tension in the string.
8 A uniform rod AB, length 1.6m, weight 30N, is freely pivoted at a fixed point A. A light elastic string BC, modulus of elasticity 60N has one end fixed at B, the other end fixed at a point C on the same horizontal level as A, AC being 2m. The system is in equilibrium with the length $BC = 1.2$m. Calculate the tension in the string, and the unstretched length of the string.

CENTRE OF MASS

CONTENTS

The centre of mass $(\bar{x}, \bar{y})$ of particles mass m_1, m_2, m_3, .. at points coordinates (x_1, y_1), (x_2, y_2), ... is given by

$$\bar{x} = \frac{m_1 x_1 + m_2 x_2 + m_3 x_3 + ...}{m_1 + m_2 + m_3 + ...}$$

i.e. $\quad \bar{x} = \frac{\Sigma m_i x_i}{\Sigma m_i},$

and $\quad \bar{y} = \frac{m_1 y_1 + m_2 y_2 + m_3 y_3 + ...}{m_1 + m_2 + m_3 + ...}$

i.e. $\quad \bar{y} = \frac{\Sigma m_i y_i}{\Sigma m_i}$

This can be written as the centre of mass of particles mass m_i at points with position vectors $x_i \mathbf{i} + y_i \mathbf{j}$ is given by

$$(\bar{x} \mathbf{i} + \bar{y} \mathbf{j}) \Sigma m_i = \Sigma m i (x_i \mathbf{i} + y_i \mathbf{j})$$

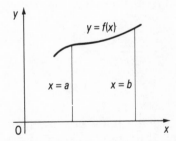

Fig 34.1

If a region R is bounded by a curve $y = f(x)$, the x-axis and the lines $x = a$, $x = b$, the coordinates of the centre of mass are given by

$$\bar{x} \int_a^b y \, dx = \int_a^b xy \, dx, \ \bar{y} \int_a^b y \, dx = \tfrac{1}{2} \int_a^b y^2 \, dx$$

If this region is rotated about the x-axis, the coordinates of the centre of mass are given by

$$\bar{x} \int_a^b y^2 \, dx = \int_a^b xy^2 \, dx, \ \bar{y} = 0$$

Similarly, if a solid of revolution is formed by rotating a similar region about the y-axis, the coordinates of the centre of mass are given by

$$\bar{x}=0 \text{ and } \bar{y}\int_c^d x^2 dy = \int_c^d yx^2 dy.$$

These results assume that the density is uniform.

CENTROID

The centre of mass of a body of uniform density is at the **centroid** of that body.

CENTRE OF GRAVITY

If we have two particles, masses m, $2m$, at points A, B respectively, then by taking moments we find that their weights are equivalent to the weight of a single particle mass $3m$ acting at a point G dividing AB in the ratio 2:1. This point G we call the centre of gravity. If these

Fig 34.2

particles are at points with displacements x_1, x_2, then the displacement of G is $\frac{1}{3}(x_1+2x_2)$. Generalizing, we find the centre of gravity of particles mass m_1, m_2, m_3, ... displacements x_1, x_2 ... is

$$\frac{m_1x_1+m_2x_2+m_3x_3+...}{m_1+m_2+m_3+...}, \text{ i.e. } \frac{\Sigma m_i x_i}{\Sigma m_i}.$$

If we have several particles in a plane, the centre of gravity $(\bar{x}, \bar{y})$ has coordinates

$$\bar{x}=\frac{\Sigma m_i x_i}{\Sigma m_i} , \bar{y}=\frac{\Sigma m_i x_i}{\Sigma m_i}$$

CENTRE OF MASS

A body only has 'weight' if it is in a gravitational field, and we find it useful to generalize the idea of a centre of gravity by defining a centre of mass to have coordinates

$$\bar{x}=\frac{\Sigma m_i x_i}{\Sigma m_i} , \bar{y}=\frac{\Sigma m_i y_i}{\Sigma m_i} \tag{1}$$

if we have a system of small bodies that we can regard as particles. If there is a uniform gravitational field, then the centre of mass coincides with the centre of gravity. the equations in (1) can be summarized in vector notation

$$\bar{x}\mathbf{i}+\bar{y}\mathbf{j}=\frac{\Sigma m_i x_i}{\Sigma m_i}\mathbf{i}+\frac{\Sigma m_i y_i}{\Sigma m_i}\mathbf{j}$$

CENTROID

We know of a point G in a triangle called the centroid, and that this point, the point of intersection of the medians, is the centre of mass of a uniform lamina covering the triangle. The centroid of any plane figure is the centre of mass of a uniform lamina covering that figure; of a body in three dimensions it is the centre of mass of a uniform solid filling that body. It is a further generalization of centre of mass, and can be defined so that it is not dependent on the idea of mass. At present we can regard centre of gravity, centre of mass and centroid as referring to the same point.

Example 34.1 Find the centre of mass of particles mass m, $2m$, $3m$, $4m$ at points with position vectors $(\mathbf{i}+\mathbf{j})$, $(2\mathbf{i}-3\mathbf{j})$, $(-\mathbf{i}+\mathbf{j})$ and $(-2\mathbf{i}-3\mathbf{j})$ respectively.

Always draw a diagram to see that your answer is reasonable.

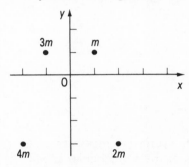

Fig 34.3

Now
$$10m(\bar{x}\mathbf{i}+\bar{y}\mathbf{j})=m(\mathbf{i}+\mathbf{j})+2m(2\mathbf{i}-3\mathbf{j})+3m(-\mathbf{i}+\mathbf{j})+4m(-2\mathbf{i}-3\mathbf{j})$$
$$=-6m\mathbf{i}-14m\mathbf{j}$$
$$\bar{x}\mathbf{i}+\bar{y}\mathbf{j}=-0.6\mathbf{i}-1.4\mathbf{j}$$

The coordinates of the centre of mass are $(-0.6, -1.4)$. From Fig 34.3, we see that this is reasonable, as the heaviest particle is in the fourth quadrant.

CENTRE OF MASS OF A ROD

A rod can be regarded as a number of small particles mass $m\delta x$ distributed along the length of the rod, where m is the mass per unit length, not necessarily uniform. To find the centre of mass we generalize equation (1) (p352) and write

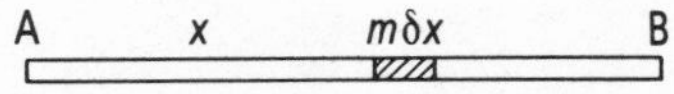

Fig 34.4

$$\bar{x}\int_0^l m\,\mathrm{d}x=\int_0^l xm\,\mathrm{d}x$$

where $\int_0^l m\,\mathrm{d}x$ is the total mass of the rod. If the mass-distribution at a point distance x from one end A is kx^2,

$$\bar{x}\int_0^l kx^2\mathrm{d}x=\int_0^l x\times kx^2\mathrm{d}x$$

$$\tfrac{1}{3}l^3 k\bar{x}=\tfrac{1}{4}kl^4$$

$$\bar{x}=\tfrac{3}{4}l$$

The centre of mass is $\tfrac{3}{4}l$ from A.

We can extend this idea into two dimensions, by dividing laminae into suitable 'rectangular strips' and regarding each strip as a particle, with mass that of the strip, placed at the 'midpoint' of the strip.

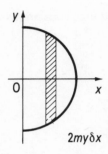

$2my\delta x$

Fig 34.5

By symmetry, the centre of mass lies on Ox (Fig 34.5). To find $\bar{x}$, we know the mass of the semicircular lamina is $\tfrac{1}{2}\pi a^2 m$, where m is the mass per unit area, so dividing the lamina into strips mass $m(2y\delta x)$,

$$\tfrac{1}{2}\pi a^2 m\bar{x}=\int_0^a x(2my)\mathrm{d}x$$

i.e. $\qquad \tfrac{1}{2}\pi a^2 \bar{x}=2\int_0^a xy\mathrm{d}x$

But the semicircle is part of the circle $x^2+y^2=a^2$ i.e. $y=\sqrt{(a^2-x^2)}$ so

$$\tfrac{1}{2}\pi a^2 \bar{x}=2\int_0^a x(a-x)^{1/2}\mathrm{d}x$$

$$=[-\tfrac{2}{3}(a^2-x^2)^{3/2}]_0^a$$

$$=\tfrac{2}{3}a^3$$

$$\bar{x}=\frac{4}{3\pi}a$$

The coordinates of the centre of mass are $((4/3\pi)a,0)$.

USE OF SYMMETRY

In Example 34.2 we used the symmetry of the lamina to help find the position of the centre of mass, and this often helps reduce the calculations required. You should form the habit of noticing any axes of symmetry that a body possesses.

A solid formed by rotating a region about an axis has that axis of rotation as an axis of symmetry, and this enables us to solve a problem in three dimensions with only one calculation.

Example 34.3 A finite region R bounded by the coordinate axes, the curve $y=e^x$ and the line $x=1$. The region is rotated completely about the x-axis. Find the centre of mass of the solid so formed when filled by matter of uniform density.

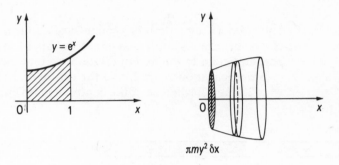

Fig 34.6

$\pi m y^2 \, \delta x$

If the region R is divided into rectangular 'strips', when these strips are rotated they form 'cylinders', mass $m\pi y^2 \delta x$, where m is the mass per unit volume of the solid. Because of the symmetry of the body, these cylinders are equivalent in this context, to particles mass $\pi y^2 \delta x$ along the axis of symmetry, so the x-coordinate of the centre of mass $\bar{x}$, is given by

$$\bar{x}\int m\pi y^2 \mathrm{d}x = \int m\pi x y^2 \mathrm{d}x$$

But $y=e^x$ is the equation of the curved boundary of R, so

$$\int_0^1 y^2 \mathrm{d}x = \int_0^1 e^{2x}\mathrm{d}x = \tfrac{1}{2}(e^2-1),$$

and

$$\int_0^1 xy^2 \mathrm{d}x = \int_0^1 x e^{2x}\mathrm{d}x$$

$$=\left[\tfrac{1}{2}xe^{2x} - \tfrac{1}{4}e^{2x}\right]_0^1$$

$$=\tfrac{1}{4}(e^2+1)$$

$$\tfrac{1}{2}(e^2-1)\bar{x}=\tfrac{1}{4}(e^2+1)$$

$$\bar{x}=\frac{e^2+1}{2(e^2-1)}$$

$$\approx 0.7$$

We expect a value a little larger than 0.5, so this is reasonable.

Example 34.4 A finite region R is bounded by one arc of the curve $y=\sin x$ and the x-axis, for $0\leqslant x\leqslant\pi$. The region is rotated completely about the x-axis. Find the centre of mass of the solid of uniform density so formed.

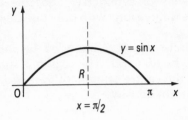

Fig 34.7

The solid of revolution is symmetrical about the x-axis, and so $y=0$. However, the region R is symmetrical about the line $x=\pi/2$, which is perpendicular to the axis of rotation, so that the resulting solid is also symmetrical about $x=\pi/2$, and the centre of mass must lie on the line $x=\pi/2$. The coordinates of the centre of mass are $(\pi/2,0)$.

QUESTIONS

Find the centre of mass of each of the following systems of particles:

1. $2m$, $3m$, $5m$ at points with position vectors $\mathbf{i}$, $3\mathbf{i}$, $5\mathbf{i}$ respectively.
2. $2m$, $3m$, $5m$ at points with position vectors $(\mathbf{i}-\mathbf{j})$, $(3\mathbf{i}+2\mathbf{j})$, $(2\mathbf{i}-5\mathbf{j})$ respectively.
3. $3m$, $7m$, $10m$ at points with position vectors $\mathbf{j}$, $(-2\mathbf{i}-5\mathbf{j})$, $(\mathbf{i}+2\mathbf{j})$ respectively.
4. m, $2m$, $2m$ at points with position vectors $\mathbf{i}$, $-\mathbf{j}$, $(2\mathbf{i}-\mathbf{j})$ respectively.
5. m, $2m$, $3m$ and $4m$ at points position vectors $\mathbf{i}$, $-3\mathbf{j}$, $(2\mathbf{i}+\mathbf{j}+\mathbf{k})$, $(-\mathbf{i}-\mathbf{j}+\mathbf{k})$ respectively.
6. Sketch each of the regions defined below, and find the coordinates of its centre of mass.
   (a) The region bounded by $y=\sin x$, and the x-axis from 0 to π.
   (b) The region bounded by the curve $y=\cos x$ and the x-axis from $-\pi/2$ to $\pi/2$.
   (c) The region bounded by $y^2=x$ and the line $x=4$.
   (d) The region bounded by $y=x^2$ and the line $y=4$.
   (e) The region bounded $y=x^{1/2}$, the x-axis and the lines $x=1$, $x=4$.
7. Sketch each of the regions described below. The region is rotated about the x-axis, to form a solid of uniform density. Use the symmetry of each body to find the coordinate of its centre of mass.
   (a) The region bounded by $y=x(2-x)$ and the x-axis between $x=0$ and $x=2$.
   (b) The region bounded by $y=x^2(2-x)^2$ and the x-axis between $x=0$ and $x=2$.

    (c)   The region bounded by $y=e^{-x^2}$ the x-axis and the lines $x=1$, $x=-1$.

    (d)   The region bounded by $y=\cos x$ and the x-axis between $\pi/2$ and $3\pi/2$.

    (e)   The region bounded by $y=\sin^2 x$ and the x-axis between 0 and π.

8   Sketch each of the regions described below. The region is rotated about the x-axis, to form a solid of uniform density. Find the coordinates of the centre of mass.

    (a) $y=x^2$, the x-axis and the line $x=2$.

    (b) $y=1/x^2$, the x-axis and the lines $x=1$, $x=2$.

9   Sketch each of the regions defined below. The region is rotated about the y-axis and forms a solid of uniform density. Find the coordinates of the centre of mass.

    (a) $y=x^2$, the y-axis and the line $y=2$.

    (b) $y=x^{1/2}$, the y-axis and the line $y=2$.

ANSWERS

CHAPTER ONE

1   Only (*a*) is one–one

2   e.g.
    (*a*) $\{x:x\geqslant 1\}$ (*b*) $\{x:x\geqslant -2\}$
    (*c*) $\left\{x:0\leqslant x\leqslant \dfrac{\pi}{2}\right\}$ (*d*) $\{x:x>0\}$

3   (*a*) $h^{-1}; y\mapsto \tfrac{1}{3}y-2; H^{-1}: y\mapsto \tfrac{1}{3}(y-2)$
    (*b*) $h^{-1}: y\mapsto \sqrt{y-1}; H^{-1}: y\mapsto \sqrt{(y-1)}$
    (*c*) $h^{-1}: y\mapsto \sqrt{\arcsin y}; H^{-1}: y\mapsto \arcsin \sqrt{y}$

4   No; odd

5   $\{y: 1\leqslant y\}, f^{-1}: y\mapsto \ln y$

6   $h:x\mapsto 2\sin x\ h^{-1}: y\mapsto \arcsin \tfrac{1}{2}y; H: x\mapsto \sin 2x, H^{-1}: y\mapsto \tfrac{1}{2}\arcsin y$

7   Even (it is a saw-tooth function).

8   Even (*b*), (*e*), (*f*); odd (*c*), (*d*); periodic (*d*) π (*e*) 2π (*f*) 2π.

CHAPTER TWO

1   (*a*) $k<1$ (*b*) $k<\tfrac{1}{3}$
    (*c*) $k<-2$ or $k>2$ (*d*) $k>-\tfrac{5}{4}$

2   (*a*)   4 (when $x=0$)   (*b*)   4 (when $x=1$)
    (*c*)   5 (when $x=1$)   (*d*)   7 (when $x=2$)

3   (*a*)   4 (when $x=0$)   (*b*)   3 (when $x=-1$)
    (*c*)   3 (when $x=-1$)   (*d*) -1 (when $x=-2$)

4   (*a*) $x=1, y=2$ or $x=2, y=-1$ (*b*) $x=1, y=2$ or $x=4, y=-7$
    (*c*) $x=1, y=2$ or $x=\tfrac{2}{3}, y=3$

5   Circle, centre origin, radius 1; not cut by the straight line.

6   $-2\sqrt{2}\leqslant k\leqslant 2\sqrt{2}$

7   $x^2+2x-2=0$

8   $x^2-3x-6=0$

9   $2x^2-4x-3=0$

10   $2x^2+4x-5=0$

11   $x^2-29x+29=0$

12   1;2

13   $k<0$ or $k>1$

14   True for all values of k

15   -1 or $\tfrac{1}{4}$

16   $x\leqslant 3, y\geqslant -1$

17   $x=\tfrac{1}{2}, y=3$

18 $\pm 9, 20$
19 $x^2 - 47x + 1 = 0$

CHAPTER THREE

1 $(x-1)(x-2)(x-7)$
2 $3x^2 - 26x + 40$; $(x-2)(3x-20)$; $(x-2)$ is a repeated factor
3 7
4 12,4
5 Other linear factors are $(x-1)(3x-2)(2x+3)$
6 (a) 8   (b) 4
  (c) $\frac{1}{4096}$ (d) $\frac{1}{16}$
  (e) $\frac{1}{2}$
7 (a) $x^{1/4}$ (b) x^{-3}
  (c) $x^{2/3}$ (d) $x^{3/2}$
  (e) x^2
8 (a) $\frac{4}{5}x^{5/4}+C$ (b) $\frac{3}{4}x^{4/3}+C$
  (c) $2x^{1/2}+C$ (d) $\frac{3}{2}x^{2/3}+C$
  (e) $\frac{3}{5}x^{5/3}+C$
9 $\frac{2}{3}(x+1)^{3/2}+\frac{2}{3}x^{3/2}+C$
10 (a) 3   (b) $\frac{1}{3}$
11 (a) $3^{1/27}$ (b) $2^{1/8}$
12 $\log_{10} x \log_5 10$; 0.4307
13 2; 1.893
14 16, -1
15 $-1, -6$
16 $-9, 0$
17 (a) 4   (b) $\frac{1}{4}$
  (c) 3.2   (d) 625
  (e) 68   (f) $\frac{5}{256}$
18 (a) $\frac{2}{5}x^{5/2}+\frac{2}{3}x^{3/2}+C$ (b) $\frac{2}{3}x^{3/2}+2x^{1/2}+C$
  (c) $\ln x - \frac{1}{x}+C$ (d) $x+\ln x+C$
19 (a) $\frac{3}{2}$ (b) $\frac{1}{6}$
  (c) $\frac{3}{4}$ (d) $\frac{4}{3}$
20 All
21 (c)
22 5
23 2

CHAPTER FIVE

1 a.s., (d); g.s., (c)
2 a.s., (a), (b), (e)
3 $-3, 105$
4 $6\frac{1}{2}, 325$

5   -1

6    $\pm\frac{1}{4}$, ±192

7   10, 22

8   201st term is 1002

9   5th term is 1250

10   7

11   (a) $8+12x+6x^2+x^3$

     (b) $1+6x+12x^2+8x^3$

     (c) $8-36x+54x^2-27x^3$

     (d) $1-12x+54x^2-108x^3+81x^4$

     (e) $16-96x+216x^2-216x^3+81x^4$

     (f) $16+96x+216x^2+216x^3+81x^4$

12   (a) $1+2x+4x^2+8x^3;\ -\frac{1}{2}<x<\frac{1}{2}$

     (b) $1+4x+12x^2+32x^3;\ -\frac{1}{2}<x<\frac{1}{2}$

     (c) $1+x+\frac{3}{2}x^2+\frac{5}{2}x^3;\ -\frac{1}{2}\leqslant x<\frac{1}{2}$

     (d) $1-x-\frac{1}{2}x^2-\frac{1}{2}x^3;\ -\frac{1}{2}\leqslant x\leqslant\frac{1}{2}$

     (e) $1+2x^2+4x^4+8x^6;\ -1\sqrt{2}<x<1/\sqrt{2}$

     (f) $\frac{1}{2}+\frac{1}{4}x+\frac{1}{8}x^2+\frac{1}{16}x^3;\ -2<x<2$

13   (a) $1-x^2+x^4,\ -1\leqslant x\leqslant1$

     (b) $1-2x^2+3x^4,\ -1<x<1$

     (c) $1-3x^2+6x^4,\ -1<x<1$

     (d) $x-2x^3+4x^5,\ -\sqrt{\frac{1}{2}}\leqslant x\leqslant\sqrt{\frac{1}{2}}$

     (e) $x+3x^3+9x^5,\ -\sqrt{\frac{1}{3}}\leqslant x\leqslant\sqrt{\frac{1}{3}}$

     (f) $x^2+x^4+x^6,\ -1\leqslant x\leqslant-\sqrt{\frac{1}{3}}$

14   (a) $\dfrac{1}{x}+\dfrac{1}{x^2}+\dfrac{1}{x^3}+\dfrac{1}{x^4}$ (b) $\dfrac{1}{x^2}-\dfrac{2}{x^4}+\dfrac{4}{x^6}-\dfrac{8}{x^8}$

     (c) $\dfrac{1}{x^2}-\dfrac{2}{x^3}+\dfrac{3}{x^4}-\dfrac{4}{x^5}$ (d) $\dfrac{1}{x^4}+\dfrac{4}{x^6}+\dfrac{12}{x^8}+\dfrac{32}{x^{10}}$

15   $1-2x-2x^2-4x^3;\ 3,\ -\frac{1}{3}$

16   (a) $-x-\dfrac{x^2}{2}-\dfrac{x^3}{3}-\dfrac{x^4}{4}$ (b) $-2x-2x^2-\dfrac{8}{3}x^3-4x^4$

     (c) $3x-\dfrac{9}{2}x^2+9x^3-\dfrac{81}{4}x^4$ (d) $3x^2-\dfrac{9}{2}x^4+9x^6-\dfrac{81}{4}x^8$

     (e) $-4x^2-8x^4-\dfrac{64}{3}x^6-64x^8$ (f) $1+2x+2x^2+\dfrac{8}{3}x^3$

     (g) $e\left(1+x+\dfrac{x^2}{2!}+\dfrac{x^3}{3!}\right)$ (h) $e\left(1+x^2+\dfrac{x^4}{2!}+\dfrac{x^6}{3!}\right)$

     (i) $2x-\dfrac{8x^3}{3!}+\dfrac{32x^5}{5!}-\dfrac{128x^7}{7!}$ (j) $1-\dfrac{x^4}{2!}+\dfrac{x^6}{3!}-\dfrac{x^8}{4!}$

17   $(1-2x)(1-x);\ -3x-\dfrac{5}{2}x^2-3x^3-\dfrac{17}{4}x^4;\ -0.328$

18   $5x-\dfrac{5}{2}x^2+\dfrac{35}{3}x^3-\dfrac{65}{4}x^4;\ \ln\left(\dfrac{1.3}{0.8}\right)=0.4855$

19   $\sin x \cos x \approx x-\dfrac{4x^3}{3!}+\dfrac{16x^5}{5!}-\dfrac{64x^7}{7!}$

20   (a) $1+\dfrac{x^2}{2!}+\dfrac{x^4}{4!}+\dfrac{x^6}{6!}$ (b) $x+\dfrac{x^3}{3!}+\dfrac{x^5}{5!}+\dfrac{x^7}{7!}$

21   2, 610

22   4; 26 terms

23   $(\tfrac{1}{2})^7$; 8

24   -12

25   5, 6

26   $-\tfrac{3}{4}x-\tfrac{139}{64}x^2$

27   $\tfrac{1}{2}, \tfrac{7}{16}, -\tfrac{35}{64}$

28   $1-2x-6x^2$

29   2, 3

30   33 952; 2.541

31   $\tfrac{1}{2}, \tfrac{1}{4}$

33   $\ln 2-\tfrac{1}{2}x-\tfrac{1}{8}x^2-\tfrac{1}{24}x^3$; $\ln 2-\tfrac{3}{2}x-\tfrac{5}{8}x^2-\tfrac{3}{8}x^3$

34   $1, \tfrac{1}{2}$

35   $-1, 1/12$

CHAPTER SIX

1   (a)   Put $x=0, -1, -2$

3   (a)   3 or -3 (b)   2 or -3

     (c)   6   (d)   0

4   (a)   Not true for any values of x;

     (b)   is an identity

5   (a)   1, -1 (b) $-1, 2$

     (c)   1, -4 (d) $-\tfrac{1}{3}, -\tfrac{1}{3}, \tfrac{1}{3}$

     (e)   1, -1 (f) $-\tfrac{1}{2}, 2, -\tfrac{3}{2}$

     (g)   1, $-1, -1$ (h) $\tfrac{1}{2}, \tfrac{1}{2}$

6   (a) $\dfrac{-7}{x-2}+\dfrac{9}{x-3}$ (b) $\dfrac{1}{x-2}-\dfrac{x}{x^2+3}$

     (c) $\dfrac{7}{(x-2)}-\dfrac{7}{(x-3)}+\dfrac{9}{(x-3)^2}$ (d) $\dfrac{-7}{5(x-2)}-\dfrac{1}{10(x+3)}+\dfrac{3}{2(x-3)}$

     (e) $2-\dfrac{11}{(x-2)}+\dfrac{21}{(x-3)}$ (f) $2(x+5)-\dfrac{16}{x-2}+\dfrac{54}{x-3}$

7   (a) $1+3x+7x^2$ (b) $x+3x^2$

     (c) $-1-2x-3x^2$ (d) $-1-3x^2$

8   (a) $\tfrac{1}{4}\ln\left|\dfrac{2x-1}{2x+1}\right|+C$ (b) $\tfrac{1}{2}\ln\left|\dfrac{x^2}{x^2+1}\right|+C$

     (c) $\tfrac{1}{2}\ln\left|\dfrac{x^2-1}{x^2}\right|+C$ (d) $\ln|x-1|-\dfrac{1}{x-1}+C$

CHAPTER SEVEN

1   (a) $x<-2$ (b) $x\leqslant4$

     (c) $x<2$ or $x>3$ (d) $-\tfrac{3}{2}\leqslant x<-1$ or $x\geqslant1$

(e) $-1<x<0$ or $x>6$

2 (a) $0<x<3$ (b) $0<x\leq1$ or $x\geq3$
   (c) $-1<x<0$ or $x>3$ (d) $x<-4$ or $-1<x<2$
3 (a) $2<x<3$ (b) $x\geq3$
4 (a) $x<-2$ or $x>2$ (b) $x<0$
   (c) $x<-6$ or $x>2$ (d) $x<-\frac{2}{3}$ or $x>8$
   (e) $x<-3$ or $x>\frac{1}{3}$
5 Curves cut the coordinate axes at
   (a) $(-4,0)$, $(2,0)$ $(0,-8)$ (b) $(-3,0)$, $(0,0)$ $(1,0)$
   (c) $(0,0)$, $(1,0)$ (d) $(-1,0)$, $(0,0)$, $(1,0)$
   (e) $(1,0)$, $(2,0)$, $(3,0)$ and $(0,6)$ (f) $(0,0)$, $(1,0)$
6 (a) $-1\leq y\leq1$ (b) $-1\leq y\leq1$
   (c) $0\leq y\leq1$ (d) $0\leq y\leq2$
   (e) $-2\leq y\leq2$
7 (a) $(0,1)$; as $x\mapsto-\infty$, $y\mapsto\infty$; as $x\mapsto\infty$, $y\mapsto\infty$
   (b) $(0,1)$; as $x\mapsto-\infty$, $y\mapsto0$, as $x\mapsto\infty$, $y\mapsto\infty$
   (c) $(1,0)$; as $x\mapsto-\infty$, $y\mapsto\infty$, as $x\mapsto\infty$, $y\mapsto\infty$ (curve is $y=2\ln|x|$)
   (d) $(1,0)$; only exists when $x>0$; as $x\mapsto\infty$, $y\mapsto\infty$
8 (a) $(0,0)$, $(\pi,0)$, $(2\pi,0)$, . . . y oscillates, increasing amplitude
   (b) $(1,0)$
   (c) $(0,0)$, $(\pi,0)$, $(2\pi,o)$, . . . y oscillates, decreasing amplitude

CHAPTER EIGHT

1   60
2   302, 400
3   2520
4   420
5   70
6   70
7   630
8   15
9   359
10 22 100; 286
11 5040; 72
12 $(2n-1)!$; $\frac{1}{2}n!(n-1)!$

CHAPTER NINE

1 (a) $\frac{9}{10}$ (b) $\frac{3}{5}$ (c) $\frac{3}{7}$
2 (a) $\frac{2}{10}$ (b) $\frac{1}{2}$ (c) $\frac{2}{7}$
3 (a) $\frac{7}{10}$ (b) $\frac{7}{10}$ (c) $\frac{4}{10}$
4 If $n\{\varepsilon\}=k$, $n\{A\}=ka$, etc. Draw Venn diagram.
5 (a)   0.1   (b)   0.45   (c) $\frac{1}{3}$
6 (a)   3/64   (b)   12/175   (c)   12/67
   (d)   12/13
7 (a)   54/175   (b)   108/175

8	(a)	0.2048	(b)	0.4872	(c)	0.2778
9	(a)	7/10	(b)	7/8	(c)	14/15
10	(a)	4/5	(b)	2/3		

11 $\frac{1}{36}, \frac{1}{6}, \frac{5}{12}; 0, \frac{1}{6}, \frac{1}{2}$

12 $\frac{1}{4}, \frac{1}{10}, \frac{9}{10}; 0, \frac{1}{10}, \frac{9}{10}$

13 0.15, 0.35, 0.075

14 $\frac{1}{36}, \frac{1}{210}, \frac{88}{105}$

15 0.72, 0.28, 0.532, 0.6192

16	(a)	0.010 24	(b)	0.0768	(c)	0.2304
	(d)	0.3456	(e)	0.2592	(f)	0.077 76
17	(a)	0.2048	(b)	0.4096	(c)	0.942 08
18	(a)	0.0250	(b)	0.0288	(c)	0.9712
19	(a)	6×10^{-3}	(b)	10^{-4}	(c)	4×10^{-2}
	(d)	1.6×10^{-3}	(e)	0.046		
20	(a)	0.028	(b)	5.9×10^{-6}	(c)	0.23
	(d)	1.4×10^{-3}				
21	(a)	0.6	(b)	0.24	(c)	0.096
	(d)	0.936				
22	(a)	0.3	(b)	0.21	(c)	0.147
23	7					
24	(a)	3/16	(b)	27/256	(c)	$(\frac{1}{4})^5$
25	6/11					
26	5/14					

CHAPTER TEN

1 Closed, not associative. No

2 Closed over **R**$^+$ but not over **Z**. Commutative but not associative.

3 Same table as cyclic group when reordered $\{1, i, -1, -i\}$

4 $1^{-1}=1, 3^{-1}=3, 5^{-1}=5, 7^{-1}$. Klein group.

5 (a) not closed (b) no unit element
  (c) not associative

6 Klein group

	f_1	f_2	f_3	f_4
f_1	f_1	f_2	f_3	f_4
f_2	f_2	f_1	f_4	f_3
f_3	f_3	f_4	f_1	f_2
f_4	f_4	f_3	f_2	f_1

7 Klein group

8 Klein group

CHAPTER ELEVEN

1 5

2 13

3    $(4, -1)$
4    $(-\frac{1}{2}, 1)$
5    $(6,5)$
6    $(10,13)$
7    $(3,1)$
8    $(0,0)$
9    $x+3y=10$
10   $y=2x-3$
11   $\frac{1}{2}x+\frac{1}{3}y=1$
12   $y=x-3$
13   $3y=2x-8$
14   $3x+2y=1$
15   $\frac{1}{7}$
16   1
17   $(x+3)^2+(y-2)^2=16$
18   $(3, -1); 5$
19   $2x^2+2y^2-10x-14y+19=0$
20   $-3x+4y=25$
21   $y=1$
22   $(4, -3); (12, -9)$
23   $x^2+y^2-3x-y=0$
24   $x^2+y^2-2x-2y+1=0; x^2+y^2-10x-10y+25=0$
25   -4 or $-\frac{56}{11}$
26   $y=mx$, where $m=\frac{1}{4}(3\pm\sqrt{3})$

CHAPTER TWELVE

1    $y^2=16x$
2    $(x-4)^2+(y-3)^2=(x+2)^2$
3    $\frac{3}{4}$
4    $y=x+2; x+y=6$
5    $am^2=ln$
6    $n^2=4lmc^2$

CHAPTER THIRTEEN

1    $(\pm\sqrt{5},0), \frac{1}{3}\sqrt{5}; (0, \pm\sqrt{5}), \frac{1}{3}\sqrt{5}$
2    $4x+3y=11, 3x-4y=2; 4x-3y=11, 3x+4y=2$
3    $y=4x\pm\sqrt{61}$
4    $y=1, x=-2; (1,1), (-5, 1), x=\frac{10}{3}, x=-\frac{22}{3}$
5    $(\pm\frac{64}{17}, \mp\frac{225}{17}); y+x=\pm\frac{161}{17}$
6    $(\pm\sqrt{13},0); x=\pm\dfrac{4}{\sqrt{13}}; 3x+2y=0(\pm\sqrt{13},0); x=\pm\dfrac{9}{\sqrt{13}}$ and $2x+3y=0$
7    $2x-y=1, x+2y=3$
8    $y=3x\pm\sqrt{11}$
9    $x=1, y=-2; (1\pm\frac{4}{5}\sqrt{41}, -2), y+2=\pm\frac{5}{4}(x-1)$
10   $(\pm\frac{25}{3}, \pm\frac{16}{3}); x+y=\pm\frac{41}{3}$

CHAPTER FOURTEEN

1   (a) $(5 \cos \theta, 5 \sin \theta)$ (b) $(\frac{1}{2} \cos \theta, \frac{1}{2} \sin \theta)$
     (c) $(t^2, 2t)$ (d) $(-t^2, 6t)$
     (e) (t, t^2)

3   $(p+q)y = 2(x + apq)$

4   (a) $\left(3t, \dfrac{3}{t}\right)$ (b) $\left(\dfrac{5t}{2}, \dfrac{5}{2t}\right)$ (c) $\left(ct, -\dfrac{c}{t}\right)$

     (d) $\left(\dfrac{-2t}{3}, \dfrac{-2}{3t}\right)$ (e) $\left(ct+1, \dfrac{c}{t}-2\right)$

5   $\left(0, \dfrac{2c}{t}\right), (2ct, 0)$

6   (a) $(3 \cos \theta, 2 \sin \theta)$ (b) $(2 \cos \theta, 5 \sin \theta)$

     (c) $\left(\dfrac{1}{\sqrt{2}} \cos \theta, \dfrac{1}{\sqrt{3}} \sin \theta\right)$ (d) (t^2, t^5)

     (e) $[t, t^2(t+1)]$

7   $\frac{1}{2}x \cos \theta + \frac{1}{3}y \sin \theta = 1$; $(2, 0)$ and $(1.6, 1.8)$

8   $\pm\sqrt{3}$

9   $y = x\sqrt{2} - 2\sqrt{2}; \dfrac{-1}{\sqrt{2}}$

CHAPTER FIFTEEN

5   (a) $r^2 \cos 2\theta = a^2$ (b) $r^2 \sin 2\theta = 2c^2$ (c) $r + 4 \cos \theta = 0$

6   (a) $xy = \frac{1}{2}$ (b) $x^2 - y^2 = 1$
     (c) $(x^2 + y^2)^2 = 2a^2 xy$

7   (a) πa^2 (b) $\frac{1}{4}\pi a^2$ (c) $\frac{1}{2}a^2$
     (d) $\frac{1}{2}a^2$ (e) $\frac{4}{3}\pi^3 a^2$

CHAPTER SIXTEEN

3   (a) $53.1°$, $(360n + 53.1)°$ or $[180(2n+1) - 53.1]°$
     (b) $154.2°$, $(360n \pm 154.2)°$
     (c) $63.4°$, $(180n + 63.4)°$
     (d) $-56.3°$, $(180n - 56.3)°$
     (e) $45.6°$, $(360n \pm 45.6)°$
     (f) $-23.6°$, $(360n - 23.6)°$ or $[180(2n+1)23.6]°$

4   (a) $\dfrac{\pi}{4}$ (b) $\dfrac{7\pi}{6}$ (c) $-\frac{1}{2}\pi$

     (d) $\dfrac{2\pi}{3}$

5   (a) 0.698 (b) 3.49 (c) -0.175
     (d) -1.75

6   (a) $\frac{\pi}{2}$, $2n\pi + \frac{\pi}{2}$ (b) $\frac{\pi}{3}$, $2n\pi \pm \frac{\pi}{3}$ (c) $\frac{\pi}{4}$, $n\pi + \frac{\pi}{4}$

7   (a)   0.305   (b)   2.21   (c)   −1.11

8   $c = 3.58$, $A = 36.4°$, $B = 98.6°$

9   $A = 46.6°$, $B = 57.9°$, $C = 75.5°$

10   1530m; 730m; 16.8°

CHAPTER SEVENTEEN

1   (a) $2 \sin 3x \cos 2x$ (b) $2 \cos 4x \cos x$ (c) $2 \cos \frac{7}{2}x \sin \frac{3}{2}x$
   (d) $2 \sin 6x \sin x$

2   (a) $\sin 8x + \sin 2x$ (b) $\cos 5x + \cos 3x$
   (c) $\frac{1}{2}(\sin 9x - \sin x)$
   (d) $-\frac{1}{2}(\cos 3x - \cos x)$

3   $8c^4 - 8c^2 + 1$, where $c \equiv \cos \theta$

4   (a)   0, 180°, 360°   (b)   60° or 120n°, $n = 0, 1, 2, 3$,
   (c)   45n°, $n = 0, 1, \ldots 8$
   (d)   0, 49.8°, 130.2°, 229.8°, 310.2°

6   (a)   −0.96   (b)   −0.8432   (c) $\frac{1}{10}\sqrt{2}$
   (d)   −0.8

7   115° or 318°

8   114° or 336°

9   40° or 252°

10   118° or 180°

11   58° or 159° or 238° or 339°

CHAPTER EIGHTEEN

1   (a) $2 + i$ (b) $5 - 4i$ (c) $2 + i$
   (d) $\frac{1}{5}(2 - i)$ (e) $\frac{1}{10}(3 + i)$

2   (a) $4 - i$ (b) $9 - 7i$ (c)   13
   (d) $\frac{1}{130}(43 + 19i)$ (e) $\frac{1}{13}(-2 + 11i)$

3   (a)   4, −3   (b)   26, 30

4   $3 - i$

5   $3 + 2i$, 13, 6

6   (a)   3, 0   (b) $2, \frac{\pi}{2}$ (c) $1, \pi$
   (d) $2, -\frac{1}{2}\pi$ (e)   5, α where $\cos \alpha : \sin \alpha : 1 = 3 : 4 : 5$

7   (a) $1, \frac{\pi}{3}$ (b) $2, -\frac{\pi}{6}$ (c) $4, \frac{5\pi}{6}$
   (d) $1, -\frac{2\pi}{3}$ (e) $2, -\frac{5\pi}{6}$

8 (a) $10(\cos \frac{1}{2}\theta + i \sin \frac{1}{2}\theta)$ (b) $\frac{2}{5}\left(\cos \frac{\pi}{6} + i \sin \frac{\pi}{6}\right)$

   (c) $30\left(\cos \frac{\pi}{4} + i \sin \frac{\pi}{4}\right)$ (d) $4\left(\cos \frac{2\pi}{3} + i \sin \frac{2\pi}{3}\right)$

   (e) $125 (\cos \frac{1}{2}\pi + i \sin \frac{1}{2}\pi)$ (f) $\frac{10}{9} (\cos \pi + i \sin \pi)$

9   $\sqrt{3} (\cos \pi/6 + i \sin \pi/6)$

   (a) $3 (\cos \pi/3 + i \sin \pi/3) = \frac{3}{2}(1 + i\sqrt{3})$

   (b) $3\sqrt{3} (\cos \pi/3 + i \sin \pi/2) = 3\sqrt{3}i$

   (c) $\frac{1}{3}(\cos \pi/3 - i \sin \pi/3) = \frac{1}{6}(\sqrt{3} - i)$

   (d) 6

10 (a) $2 \cos \frac{1}{2}\theta(\cos \frac{1}{2}\theta + i \sin \frac{1}{2}\theta)$ (b) $4 \cos^2 \frac{1}{2}\theta(\cos \theta + i \sin \theta)$

   (c) $\dfrac{1}{2 \cos \frac{1}{2}\theta} [\cos (-\frac{1}{2}\theta) + i \sin (-\frac{1}{2}\theta)]$

                                   (d) $16 \cos^4 \frac{1}{2}\theta (\cos 2\theta + i \sin 2\theta)$

11   Circle centre (0, 0), radius 2; centre (2, 3), radius 2

15   (1, 1)

17   $2i, \sqrt{3} - i$

18   $-1 + i, 2 + i$

20   $\sqrt{2} - i\sqrt{2}, -\sqrt{2} \pm i\sqrt{2}$

21   Circle $|z| = 2$, described in clockwise sense; circle $|z| = \frac{1}{2}$, described in clockwise sense

24   $32c^6 - 48c^4 + 18c^2 - 1 = 0$, where $c = \cos \theta$

CHAPTER NINETEEN

1 (a)   Enlargement factor 3

   (b)   Reflection in y-axis

   (c)   Reflection in $y = x$

   (d)   Rotation about O through $+90°$ i.e. anti-clockwise

   (e)   Rotation about O through $-90°$

3 (a) $y = 2x$ (b) $x = 0$ (c) $y = 2x$

4 (a) $y = -x$ (b) $y = -\frac{1}{2}x$ (c) $y = -3x$

5 (a) $-72, -36, -4, -4$ (b) $-72, 36, -4, 4$

6   $\begin{pmatrix} -72 & 56 & -8 \\ 36 & -28 & 5 \\ -4 & 4 & -1 \end{pmatrix}$

7   $\begin{pmatrix} -9 & 7 & -1 \\ \frac{9}{2} & -\frac{7}{2} & \frac{5}{8} \\ -\frac{1}{2} & \frac{1}{2} & -\frac{1}{8} \end{pmatrix}$

8   8

9 (a)   Singular   (b) $\begin{pmatrix} -3 & -2 & 3 \\ 2 & 4 & -3 \\ 2 & -1 & -1 \end{pmatrix}$

10   Maps plane $x + y = 0$ into the origin

11 (a) $3x = 4y$ (b) $4x - 3y = 0$

   (c)   Every point on $3y = x$ maps on to itself, the line $y = -3x$ maps on to itself

12 5

13 $\dfrac{1}{7}\begin{pmatrix} 5 & 2 & -1 \\ -3 & 3 & 2 \\ -3 & -4 & 2 \end{pmatrix}$; 2, 1, -3

CHAPTER TWENTY

1 (b) $\begin{pmatrix} 8 \\ 9 \end{pmatrix}$ (c) $\begin{pmatrix} 5\frac{1}{2} \\ 8 \end{pmatrix}, \begin{pmatrix} 7 \\ 6\frac{1}{2} \end{pmatrix}$ (d) $\begin{pmatrix} 1\frac{1}{2} \\ -1\frac{1}{2} \end{pmatrix}$

2 (a) $\begin{pmatrix} -4 \\ -5 \end{pmatrix}$ (b) $\begin{pmatrix} 2 \\ 3 \end{pmatrix}$

3 (a) $\begin{pmatrix} 3\frac{1}{2} \\ 4\frac{1}{2} \end{pmatrix}$ (b) $\begin{pmatrix} 2 \\ 1 \end{pmatrix}$ (c) $\begin{pmatrix} 2 \\ 2 \end{pmatrix}$

4 $\sqrt{6}, 3\sqrt{6}, 3\sqrt{2}$

5 $\begin{pmatrix} 4 \\ 7 \\ 6 \end{pmatrix}, \begin{pmatrix} 0 \\ 1 \\ 1 \end{pmatrix}, \begin{pmatrix} 6 \\ 10 \\ 7 \end{pmatrix}$

6 (a) $\mathbf{r}=\begin{pmatrix} 2 \\ 1 \end{pmatrix}+\mathbf{t}\begin{pmatrix} -1 \\ -1 \end{pmatrix}$ (b) $\mathbf{r}=\begin{pmatrix} 2 \\ 1 \end{pmatrix}+t\begin{pmatrix} -6 \\ 1 \end{pmatrix}$ (c) $\mathbf{r}=\begin{pmatrix} 2 \\ 1 \end{pmatrix}+t\begin{pmatrix} 0 \\ 1 \end{pmatrix}$

7 A and C

8 $\mathbf{r}=\begin{pmatrix} 1 \\ 3 \end{pmatrix}+\mathbf{s}\begin{pmatrix} 2 \\ 4 \end{pmatrix}, y=2x+1$

(b) $\mathbf{r}=\begin{pmatrix} 3 \\ 7 \end{pmatrix}+t\begin{pmatrix} 2 \\ -8 \end{pmatrix}, y+4x=19$

(c) $\mathbf{r}=\begin{pmatrix} 5 \\ -1 \end{pmatrix}+t\begin{pmatrix} 4 \\ -4 \end{pmatrix}, x+y=4$

9 $\mathbf{r}=\begin{pmatrix} 3 \\ 1 \\ 2 \end{pmatrix}+t\begin{pmatrix} -1 \\ 2 \\ -3 \end{pmatrix}$

10 $\mathbf{r}=\begin{pmatrix} 2 \\ 1 \\ 0 \end{pmatrix}+t\begin{pmatrix} 2 \\ -1 \\ -1 \end{pmatrix}, \dfrac{x-2}{2}=\dfrac{y-1}{-1}=\dfrac{2}{-1}$

$\mathbf{r}=\begin{pmatrix} 4 \\ 0 \\ -1 \end{pmatrix}+t\begin{pmatrix} 1 \\ -1 \\ 2 \end{pmatrix}, \dfrac{x-4}{1}=\dfrac{y}{-1}=\dfrac{z+1}{2}$

$\mathbf{r}=\begin{pmatrix} 2 \\ 1 \\ 0 \end{pmatrix}+t\begin{pmatrix} 3 \\ -2 \\ 1 \end{pmatrix}, \dfrac{x-2}{3}=\dfrac{y-1}{-z}=\dfrac{z}{1}$

11 4, -9, 4

13 $1/\sqrt{10}$

14 $-\dfrac{6}{5\sqrt{13}}, -\dfrac{1}{5\sqrt{2}}, \dfrac{5}{\sqrt{26}}$

15 $0.8\mathbf{i}+0.6\mathbf{j}$

16 $\pm\dfrac{1}{\sqrt{2}}(\mathbf{i}-\mathbf{k})$

18 $-\dfrac{\sqrt{3}}{2\sqrt{7}}$

19 $-\dfrac{1}{5\sqrt{2}}$

20 $\pm\dfrac{1}{\sqrt{5}}(\mathbf{i}+2\mathbf{j})$; e.g. **k**

21 (a)　$x-2y+3z=4$　(b)　$2x-3y=12$　(c)　$x=3$
22 (a)　$\mathbf{i}+2\mathbf{j}+2\mathbf{k}$　(b)　$4\mathbf{i}+\mathbf{j}+\mathbf{k}$　(c)　$\mathbf{i}+\mathbf{j}+3\mathbf{k}$
23 (a)　$\mathbf{i}+2\mathbf{j}+2\mathbf{k}$　(b)　$\mathbf{r}=3\mathbf{i}-2\mathbf{j}+\mathbf{k}+t(\mathbf{i}+2\mathbf{j}+2\mathbf{k})$
　　(c)　$(4, 0, 1)$　(d)　3　(e)　$5\mathbf{i}+2\mathbf{j}+3\mathbf{k}$
24 (a)　$3, -2\mathbf{i}+\mathbf{j}+8\mathbf{k}$　(b)　$\sqrt{5}, 5\mathbf{i}-4\mathbf{j}+\mathbf{k}$　(c)　$\sqrt{14}, 9\mathbf{i}$

25 $-\dfrac{1}{3\sqrt{3}}$

26 (a)　Several possible forms, one of which is $\mathbf{r}=2\mathbf{i}+\mathbf{j} + s(-\mathbf{i}+\mathbf{k}) + t(\mathbf{i}-2\mathbf{j}-\mathbf{k})$
　　(b)　$\mathbf{r}.(3\mathbf{i}+2\mathbf{j}+\mathbf{k})=6$
27 $\mathbf{r}=4\mathbf{i}-3\mathbf{j}-2\mathbf{k}+s(6\mathbf{i}-4\mathbf{j}-4\mathbf{k})+t(\mathbf{i}-2\mathbf{j}-\mathbf{k})$ $\mathbf{r}.(2\mathbf{i}-\mathbf{j}+4\mathbf{k})=3$
28 $3x+2y-z=5$; $\mathbf{r}=\mathbf{i}+2\mathbf{j}+2\mathbf{k}+s(\mathbf{j}+2\mathbf{k})+t(\mathbf{i}+3\mathbf{k})$
29 $4x+3y+2z=3$
30 The plane is $\mathbf{r}.(3\mathbf{i}+5\mathbf{j}+\mathbf{k})=2$
31 (a)　$\mathbf{a}-\mathbf{b}+\mathbf{c}$
　　(b)　$\mathbf{c}+\mathbf{b}-\mathbf{a}$
32 $3\mathbf{i}+2\mathbf{j}; \sqrt{0.6}$
34 $(\mathbf{r}-\mathbf{i}-\mathbf{j}).(\mathbf{r}-3\mathbf{i}+5\mathbf{j})=0$; $(\mathbf{r}-3\mathbf{i}+4\mathbf{j}-\mathbf{k}).(\mathbf{r}-\mathbf{i}+2\mathbf{j}-3\mathbf{k})=0$
35 $\frac{5}{2}\sqrt{5}, \frac{1}{2}\sqrt{38}$
36 (a)　$5\mathbf{i}-\mathbf{k}$　(b)　$(5, 0, -1)$, same plane and same line
37 $\dfrac{8}{7\sqrt{6}}$

38 $\dfrac{4}{\sqrt{406}}$

CHAPTER TWENTY-ONE

1 (a)　$\dfrac{x}{\sqrt{(1+x^2)}}$　(b)　$\dfrac{-2}{(1+2x)^2}$　(c)　$4\cos 4x$

　　(d)　$\dfrac{1}{x}$　(e)　$3e^{3x+2}$

2 (a)　$\dfrac{-3}{(3x+2)^2}$　(b)　$2\sin x\cos x$　(c)　$-\tan x$

　　(d)　$\cos x\,e^{\sin x}$　(e)　$\dfrac{\cos x}{2\sqrt{(1+\sin x)}}$

3 (a)　$\frac{2}{3}$　(b)　$-\frac{2}{9}\sqrt{3}$　(c)　1
4 $\sin 3x+3x\cos 3x$
5 $2x\cos 2x-2x^2\sin 2x$

6    $x^3 \sec^2 x + 3x^2 \tan x$

7    $e^{2x}(1+2x)$

8    $2x(1+x)e^{2x+1}$

9    $1+\ln x$

10   $\dfrac{x(5x+4)}{2(x+1)^{\frac{1}{2}}}$

11   $\dfrac{x \cos x - \sin x}{x^2}$

12   $-\dfrac{x \sin x + 2 \cos x}{x^3}$

13   $\dfrac{\cos x - \sin x}{e^x}$

14   $\dfrac{1-2\ln x}{x^3}$

15   $-\dfrac{2 \sin x \sin 2x + \cos x \cos 2x}{\sin^2 x}$

16   (a) $\dfrac{-\sin y}{1+x \cos y}$ (b) $\dfrac{\cos x - y}{x+3y^2}$ (c) $\dfrac{2x-ye^x}{2y+e^x}$

17   -1

18   (a) $\frac{3}{2}t$ (b) $\dfrac{-\sin t}{2 \cos 2t}$ (c) -1

19   $\left(\dfrac{1-t^2}{1+t^2}\right)^3$

20   $\dfrac{-2 \sin 2x}{1+\cos 2x}$

21   (a) $\dfrac{2}{x}$ (b) $\dfrac{2}{x}\ln x$

22   $1-\dfrac{x \arcsin x}{\sqrt{(1-x^2)}}$

23   $\dfrac{-\sin t}{1+\cos t}$; $x+y=2+\dfrac{\pi}{2}$

24   (a) $\dfrac{4x}{(x^2+1)^2}$ (b) $1+\dfrac{1}{(x+1)^2}$

25   $\dfrac{1-x^2}{x(1+x^2)}$

26   $e^{-x}(1-x^2)$

CHAPTER TWENTY-TWO

1   (a) $(0, 0)$, point of inflexion

    (b) $(-1, 0)$ and $(1, 0)$, minima; $(0,1)$ maximum

    (c) $(-\frac{1}{4}, -\frac{1}{256})$, minimum; $(0, 0)$, point of inflexion

    (d $\left(-\dfrac{2}{3}, \dfrac{-2\sqrt{3}}{9}\right)$, minimum

2   (a)   min. value 3 when $x=1$

     (b)   max. value 0 when $x=0$, min. value $-\frac{4}{27}$ when $x=\frac{2}{3}$

     (c)   max. value $\dfrac{2\sqrt{3}}{9}$ when $\cos x=\frac{1}{3}\sqrt{3}$

     (d)   max. value $e^{-\frac{1}{2}}$ when $x=1$; min. value $-e^{-\frac{1}{2}}$ when $x=-1$

3   100

4   5

5   $4\sqrt{3}$

6   25cm^2

7   3.75%; 3.8%

8   4%

9   $(1, \frac{1}{2})$, max; $(-1, -\frac{1}{2})$, min

10   $(0, 1)$, max; $(-1, 0)$, min

11   $\frac{1}{3}$; $(2e-3)/e^2$

12   Radius of circle centre $(0, 0)$ touching $3x+4y=5$ is 1.

13   (a)   1%

     (b)   6.4%

16   $\left(\dfrac{\pi}{6}, \frac{1}{3}\sqrt{3}\right)$, max; $\left(\dfrac{5\pi}{6}, -\frac{1}{3}\sqrt{3}\right)$, min

18   3.5 metres

CHAPTER TWENTY-THREE

1   $\frac{1}{2}x+\frac{1}{4}\sin 2x+C$

2   $\frac{1}{4}\sin 2x+\frac{1}{8}\sin 4x+C$

3   $-\frac{1}{10}\cos 5x-\frac{1}{2}\cos x+C$

4   $\frac{1}{8}\sin 4x-\frac{1}{12}\sin 6x+C$

5   $\dfrac{\pi}{2}$

6   $\frac{1}{2}\pi$

7   $\frac{2}{3}-\frac{1}{4}\sqrt{3}$

8   $\frac{1}{3}$

9   0

10   $\frac{3}{16}\pi$

11   $\ln (x+2)+C$

12   $x-2\ln (x+2)+C$

13   $\frac{1}{2}x^2-2x+4\ln (x+2)+C$

14   $\ln (2-\cos x)+C$

15   $\frac{1}{2}\ln (1+e^{2x})+C$

16   $\frac{1}{3}\ln 2$

17   $\frac{1}{3}\ln (\frac{8}{5})$

18   $\ln 2$

19   $\ln \frac{1}{2}(\sqrt{3}-1)$

20   $\frac{1}{2}\ln (\frac{8}{5})$

21   $\arc \sin \left(\dfrac{x}{2}\right)+C$

22    $\frac{1}{2}\arctan\left(\frac{x-1}{2}\right)+C$

23    $\frac{1}{2}\arctan 2x+C$

24    $\frac{1}{2}\arcsin 2x+C$

25    $\dfrac{1}{15}\arctan\left(\dfrac{3x}{5}\right)+C$

26    $\dfrac{\pi}{6}$

27    $\dfrac{\pi}{12}$

28    $\dfrac{\pi}{6}$

29    $\dfrac{\pi}{6}$

30    $\dfrac{\pi}{20}$

31    $\arctan(x+2)+C$

32    $\frac{1}{2}\ln\left(\dfrac{x+1}{x+3}\right)+C$

33    $\frac{116}{15}$

34    $\dfrac{\pi}{4}$

35    $\dfrac{\pi}{2}$

36    $-x\cos x+\sin x+C$

37   xe^x-e^x+C

38    $e^x(x^2-2x+2)+C$

39    $\frac{1}{3}x^3(\ln x-\frac{1}{3})+C$

40    $x\arctan x-\frac{1}{2}\ln(1+x^2)+C$

41    $\frac{4}{3}; \frac{16}{15}\pi$

42    $\frac{4}{3}, \frac{56}{15}\pi$

43    $2; \dfrac{\pi^2}{2}$

44    $\frac{5}{24}\pi a^3$

45    $\dfrac{\pi}{2}\left(e^4-1\right)$

46    $\frac{1}{4}$

47   30

48    $\dfrac{2}{\pi}$

49   0

50    $\frac{1}{2}$

51   (a) $-2(4-x)^{\frac{1}{2}}+C$ (b) $-\ln(4-x)+C$

       (c) $\arcsin\left(\dfrac{x}{2}\right)+C$ (d) $\frac{1}{2}\arctan\left(\dfrac{x}{2}\right)+C$

52 $\text{arc tan } (e^x) + C$

53 $x \text{ arc sin } x + (1-x^2)^{\frac{1}{2}} + C$

54 $1 + \frac{1}{2} \ln 2 - \frac{9}{2} \ln (\frac{4}{3})$

55 $2(\sqrt{2}-1); \pi \ln 2; \frac{4}{3}\pi(2-\sqrt{2})$

58 $\frac{1}{2} \ln \dfrac{x^2(1+x)}{(1-x)} + C$

59 $\dfrac{4}{3\pi}$

60 $\dfrac{2}{\pi}; \dfrac{\pi}{4}$

CHAPTER TWENTY-FOUR

1 $y = x^3 + x^4 - 1$

2 $y = -\cos x + e^x$

3 $y = \frac{1}{3}x^3 - e^{-x} + 1$

4 $y = \frac{1}{2} \sin 2x - \frac{1}{2}$

5 $y = 2e^{5x}$

6 $y = 2e^{3x}$

7 $y = 5e^{-4x}$

8 $y = e^{(2-x)}$

9 $\dfrac{1}{y} + \ln x - 1 = 0$

10 $\cos y = x - 1$

11 $\text{arc tan } y = \dfrac{\pi}{4} + \ln \sec x$

12 $y - \dfrac{1-2\cos^2 x}{1+2\cos^2 x}$

13 $\dfrac{1}{y} = 2 - \ln x - x^2$

14 $y = Ae^x - 1 - x$

15 $y = \tan\left(x - \dfrac{\pi}{4}\right) - x$

16 $y^2 = 2x^2(\ln x + \frac{1}{2})$

17 $y^2 = \frac{1}{2}(x^2 - x^{-2})$

18 $y = e^{2x} + e^{-2x}$

19 $y = 2 \cos 2x$

20 $y = e^{3x} - e^{-3x}$

21 $y = 2 \sin 3x$

22 $y = \sin x + 3 \cos x$

23 $y = 2e^{2x} - e^{-x}$

24 $y = 3 \sin \frac{1}{2}x$

25 $y = \frac{3}{2}[e^{\frac{1}{2}(x-1)} + e^{-\frac{1}{2}(x-1)}]$

26 $y = 5[\cos(\frac{2}{5}x) + \frac{1}{2} \sin(\frac{2}{5}x)]$

27 $y = 5e^{2x/5}$

28 $y = 2 + \cos 2x$

29 $y = 3e^{2x} + 3e^{-2x} - 2$

30    $y=xe^x$

31    $y^4=1+2x^2$

32    $\dfrac{1}{y^2}=x^2-2x+2$

33    $\ln y=\tfrac{1}{4}x^2-1$

34    $3y+x^3+3x^4=19$

35    $y=Ae^{-4x}+B$

36    $y=2x+B+Ae^{-4x}$

37    $y(1+Ax^2)=2$

38    $y=\dfrac{2}{(1-Ae^{\frac{1}{2}x^2})}$

39    $y(A\cos x-1)=1$

CHAPTER TWENTY-FIVE

1   0.6938; 0.693 15; 0.693 15

2   1.2910; 1,881

3   54.5; 54.3

6    $1+x+\tfrac{1}{2}x^2-\tfrac{1}{8}x^4$

7    $x+x^2+\tfrac{1}{3}x^3$

9   0.414

10   0.167

11   2.201

12   1.972

13   1.21

14   4.56

15   0.3398; $F'(2)=6>1$; $x_{r+1}=(6x_r-2)^{\frac{1}{3}}$

16    $F(x)=\tfrac{1}{10}(x^3+1)$ gives $x\approx0.10$; $F(x)=(10x-1)^{1/3}$ gives $x\approx3.11$ and $x\approx-3.21$

17   1.17

18    $x_{r+}=\ln(6/x_r)$; 1.43

CHAPTER TWENTY-SIX

1   14m s^{-1}; 60m

2   10m s^{-1}; $\tfrac{1}{9}\text{m s}^{-2}$

3   3.6km

4   6.5 s

5   7.5 minutes

6   48m s^{-1}; 64m, 35m

7   11.25m s^{-1}; 187.5m

8   2m s^{-1}

9   1.26m s^{-1}; 0.74m

10   3m s^{-1}; 4m s^{-1}

11   0.58m, 6m

12   12.6m s^{-1}; 2m

13   31.25m s^{-1}, 75m s^{-1}; 937.5W, 3kW

14   200N

15   1500N; 25.3kW; 15.5kW; 9.1kW

16   710m

CHAPTER TWENTY-SEVEN

1 $2t\mathbf{i}+3t^2\mathbf{j}$; $2\mathbf{i}+6t\mathbf{j}$; $\sqrt{160}$; $8\mathbf{i}+240\mathbf{j}$

3 $-e^{-t}\mathbf{i}+2t\mathbf{j}$; $e^{-t}\mathbf{i}+2\mathbf{j}$; $2(e^{-t}\mathbf{i}+2\mathbf{j})$

4 $4\mathbf{i}-4\mathbf{j}+10\mathbf{k}$

5 $\mathbf{v}=t^2\mathbf{i}-4t^3\mathbf{j}$; $\mathbf{s}=\frac{1}{3}t^3\mathbf{i}-t^4\mathbf{j}$

6 $-\mathbf{i}+k\mathbf{j}+t(\mathbf{i}-2\mathbf{j})$; $k=2$

7 $\sqrt{(5t^2-16t+13)}$; $t=\frac{8}{5}$

8 $6\mathbf{i}-2\mathbf{j}$; $12\mathbf{i}$; $-10\mathbf{i}-5\mathbf{j}$

9   10, 3

10 $m(t^2\mathbf{i}+t\mathbf{j}+\mathbf{k})$; $\frac{1}{2}m(t^4+t^2+1)$; $m(2t\mathbf{i}+\mathbf{j})$; $mt(2t^2+1)$

12 $\dfrac{17}{\sqrt{410}}$

13 $12(\mathbf{i}+\mathbf{j})$m s^{-1}, $24(\mathbf{i}+\mathbf{j})$ Ns, 288 J

14 $5\mathbf{i}+2\mathbf{j}$

15 $3\mathbf{i}+4\mathbf{j}-3\mathbf{k}$

16 $4\mathbf{i}+6\mathbf{j}$

17 $\frac{1}{11}$

18   100 J

19 $\frac{1}{2}(7\mathbf{i}-\mathbf{j})$m s^{-2}; $\frac{5}{2}(7\mathbf{i}-\mathbf{j})$m s^{-1}, 225 J, 250 W

20   75 J

21 $\dfrac{1}{1+t^2}$

22 $(8\mathbf{i}+16\mathbf{j})$ Ns, 80 J, 304 W

CHAPTER TWENTY-EIGHT

1 $\frac{1}{3}$

2 $3mu^2$; $\frac{9}{4}mu^2$

3 $\frac{1}{3}$; 0

4 $\frac{1}{2}$; $\frac{2}{3}$

5 $\sqrt{\frac{2}{3}}$

6 $mu(1+e)$

7 $1.2\mathbf{i}$, $2.4\mathbf{i}$; 2.8N s; 2.52 J

CHAPTER TWENTY-NINE

1   125m; 26.6° or 63.4°; 12.5m, 50m

2 $\dfrac{2uv}{g}$

5   80° or −58°; 58° or −71°; 3:1

6    $\dfrac{\pi}{12}$ or $\dfrac{5\pi}{12}$

CHAPTER THIRTY

1    $4\pi^2 m^2 Ml$; $\arccos\left(\dfrac{g}{4\pi^2 m^2 l}\right)$

2    $\frac{1}{9}mg$; $\frac{3}{2}\sqrt{ga}$

3   0.2m

4    $\frac{1}{3}l$

5    $\frac{1}{3}$

CHAPTER THIRTY-ONE

1    $80\,\text{Nm}^{-1}$; $1.6\times10^3\,\text{Nm}^{-1}$

2   160N, 400N

3   1N

4   0.5N

5   c; $\frac{1}{2}c$; $2c$

6   0.2 J, 0.6 J, 0.7 J

7    $m\ddot{x}=mg-\dfrac{4mgx}{a}$, i.e. $\ddot{x}+\dfrac{4g}{a}\left(x-\tfrac{1}{4}a\right)=0$; s.h.m. about $x=\tfrac{1}{4}a$; $\pi\sqrt{\dfrac{a}{g}}$

8    $d\sqrt{\left(\dfrac{\lambda}{a(M+m)}\right)}$; $m\left(\dfrac{d\lambda}{g+a(M+m)}\right)$

CHAPTER THIRTY-TWO

1   After 1 s; $5.8\,\text{m s}^{-1}$

3    $\dfrac{u}{g(\sin\alpha+\mu\cos\alpha)}$; $\dfrac{u^2}{2g(\sin\alpha+\mu\cos\alpha)}$; $\dfrac{u}{g\sqrt{(\sin^2\alpha-\mu^2\cos^2\alpha)}}$

CHAPTER THIRTY-THREE

1   7.4 N

2   5.2 N

3   8.9 N

4   364 N, 0.3

5   (a)   880 N, 264 N

        (b) $40(1+20x)\cos 70°$

6   43 N

7   70 N

8   12 N, 1m

1   3.6**i**

2   2.1**i**−2.1**j**

3   −0.2**i**−0.6**j**

4   **i**−0.8**j**

5   0.3**i**−0.7**j**+0.7**k**

6   (a) $\left(\dfrac{\pi}{2}, \dfrac{\pi}{8}\right)$ (b) $\left(0, \dfrac{\pi}{8}\right)$ (c)   (2.4, 0)

   (d)   (0, 2.4)   (e) $\left(\dfrac{93}{35}\right), \left(\dfrac{45}{56}\right)$

7   (a)   (1, 0)   (b)   (1, 0)   (c)   (0, 0)

   (d)   (π, 0)   (e) $\left(\dfrac{\pi}{2}, 0\right)$

8   (a) $\left(\tfrac{5}{3}, 0\right)$ (b) $\left(\tfrac{9}{7}, 0\right)$

9   (a) $\left(0, \tfrac{4}{3}\right)$ (b) $\left(0, \tfrac{5}{3}\right)$

INDEX

More advanced level exam help from Pan

BRODIE'S NOTES
on English Literature texts

This popular and respected series provides reliable guidance on texts commonly set for literature exams in the UK and the Republic of Ireland.

Brodie's Notes on texts set for A level, Highers and Leaving Certificate reflect the deeper critical appreciation and analysis required for study at such levels.

Some selected Brodie's Notes for advanced level are:

William Shakespeare
ANTONY AND CLEOPATRA
CORIOLANUS
HAMLET
KING LEAR
OTHELLO
THE TEMPEST

Geoffrey Chaucer
THE MILLER'S TALE†
THE NUN'S PRIEST'S
   TALE†
THE PARDONER'S TALE†
THE WIFE OF BATH'S TALE†
(With parallel texts)

Jane Austen
EMMA
MANSFIELD PARK
SENSE AND SENSIBILITY

Emily Brontë WUTHERING HEIGHTS
Joseph Conrad HEART OF DARKNESS
T. S. Eliot MURDER IN THE CATHEDRAL
Thomas Hardy RETURN OF THE NATIVE
TESS OF THE D'URBERVILLES
James Joyce DUBLINERS
Christopher Marlowe DR FAUSTUS
Thomas Middleton
& William Rowley THE CHANGELING
John Milton COMUS/SAMSON AGONISTES
Tom Stoppard ROSENCRANTZ AND GUILDENSTERN ARE
   DEAD
Jonathan Swift GULLIVER'S TRAVELS
John Webster THE DUCHESS OF MALFI
THE WHITE DEVIL
William Wordsworth THE PRELUDE Books 1 and 2
William Wycherley THE COUNTRY WIFE

Various THE METAPHYSICAL POETS

*Written by experienced teachers and examiners who can give you effective advice

*Designed to increase your understanding, appreciation and enjoyment of a set work or author

*With textual notes, commentaries, critical analysis, background information and revision questions

£1.25 each except for †£1.75. On sale in bookshops.

For information write to:

Pan Study Aids & Brodie's Notes
Pan Books Ltd
18–21 Cavaye Place
London SW10 9PG

For business courses at school or college

BREAKTHROUGH BUSINESS BOOKS

This series covers a wide range of subjects for students following business and professional training syllabuses. The Breakthrough books make ideal texts for BTEC, SCOTVEC, LCCI and RSA exams.

Many business teachers and lecturers have praised the books for their *excellent value, clear presentation, practical, down-to-earth style* and *modern approach to learning*.

'An excellent range of texts at a price students can afford.'
'The self-study presentation and style make these ideal college books.'

The range of 30 includes the following major titles:

BACKGROUND TO BUSINESS	£3.50
BUSINESS ADMINISTRATION	
A fresh approach	£3.95
THE BUSINESS OF COMMUNICATING	£3.95
WHAT DO YOU MEAN 'COMMUNICATION'	£3.95
THE ECONOMICS OF BUSINESS	£2.95
EFFECTIVE ADVERTISING AND PR	£2.95
MANAGEMENT	
A fresh approach	£3.95
MARKETING	
A fresh approach	£3.50
PRACTICAL COST AND MANAGEMENT ACCOUNTING	£3.50
UNDERSTANDING COMPANY ACCOUNTS	£2.95
PRACTICAL BUSINESS LAW	£3.95

On sale in bookshops.

For information write to:

Business Books
Pan Books Ltd
18–21 Cavaye Place
London SW10 9PG